Betti Numbers of the Moduli Space of Rank 3 Parabolic Higgs Bundles

MEMOIRS
of the
American Mathematical Society

Number 879

Betti Numbers of the Moduli Space of Rank 3 Parabolic Higgs Bundles

O. García-Prada
P. B. Gothen
V. Muñoz

May 2007 • Volume 187 • Number 879 (end of volume) • ISSN 0065-9266

American Mathematical Society
Providence, Rhode Island

2000 *Mathematics Subject Classification.* Primary 14D20, 14H60.

Library of Congress Cataloging-in-Publication Data

García-Prada, O. (Oscar), 1960
  Betti numbers of the moduli space of rank 3 parabolic Higgs bundles / O. García-Prada, P. B. Gothen, V. Muñoz.
    p. cm. — (Memoirs of the American Mathematical Society, ISSN 0065-9266 ; no. 879)
  "May 2007, volume 187, number 879 (end of volume)."
  Includes bibliographical references.
  ISBN 978-0-8218-3972-0 (alk. paper)
  1. Vector bundles. 2. Moduli theory. I. Gothen, P. B. (Peter Beier), 1967– II. Muñoz, V. (Vicente), 1971– III. Title.
QA612.63.G37 2007
514'.224—dc22   2007060665

Memoirs of the American Mathematical Society

This journal is devoted entirely to research in pure and applied mathematics.

Subscription information. The 2007 subscription begins with volume 185 and consists of six mailings, each containing one or more numbers. Subscription prices for 2007 are US$649 list, US$519 institutional member. A late charge of 10% of the subscription price will be imposed on orders received from nonmembers after January 1 of the subscription year. Subscribers outside the United States and India must pay a postage surcharge of US$38; subscribers in India must pay a postage surcharge of US$43. Expedited delivery to destinations in North America US$53; elsewhere US$130. Each number may be ordered separately; *please specify number* when ordering an individual number. For prices and titles of recently released numbers, see the New Publications sections of the *Notices of the American Mathematical Society*.

Back number information. For back issues see the *AMS Catalog of Publications*.

Subscriptions and orders should be addressed to the American Mathematical Society, P. O. Box 845904, Boston, MA 02284-5904, USA. *All orders must be accompanied by payment.* Other correspondence should be addressed to 201 Charles Street, Providence, RI 02904-2294, USA.

Copying and reprinting. Individual readers of this publication, and nonprofit libraries acting for them, are permitted to make fair use of the material, such as to copy a chapter for use in teaching or research. Permission is granted to quote brief passages from this publication in reviews, provided the customary acknowledgment of the source is given.

Republication, systematic copying, or multiple reproduction of any material in this publication is permitted only under license from the American Mathematical Society. Requests for such permission should be addressed to the Acquisitions Department, American Mathematical Society, 201 Charles Street, Providence, Rhode Island 02904-2294, USA. Requests can also be made by e-mail to reprint-permission@ams.org.

Memoirs of the American Mathematical Society is published bimonthly (each volume consisting usually of more than one number) by the American Mathematical Society at 201 Charles Street, Providence, RI 02904-2294, USA. Periodicals postage paid at Providence, RI. Postmaster: Send address changes to Memoirs, American Mathematical Society, 201 Charles Street, Providence, RI 02904-2294, USA.

© 2007 by the American Mathematical Society. All rights reserved.
Copyright of this publication reverts to the public domain 28 years
after publication. Contact the AMS for copyright status.
This publication is indexed in *Science Citation Index*®, *SciSearch*®, *Research Alert*®, *CompuMath Citation Index*®, *Current Contents*®/*Physical, Chemical & Earth Sciences*.
Printed in the United States of America.

∞ The paper used in this book is acid-free and falls within the guidelines
established to ensure permanence and durability.
Visit the AMS home page at http://www.ams.org/

10 9 8 7 6 5 4 3 2 1   12 11 10 09 08 07

Contents

Chapter 1. Introduction — 1

Chapter 2. Parabolic Higgs bundles — 5
  1. Definitions and basic facts — 5
  2. Deformation theory — 7
  3. Parabolic Higgs bundles and gauge theory — 9

Chapter 3. Morse theory on the moduli space — 11
  1. The Morse function — 11
  2. Fixed points of the S^1 action on the moduli space — 12
  3. Morse indices — 14
  4. Rank three parabolic Higgs bundles — 16
  5. Laumon's Theorem for parabolic Higgs bundles — 16

Chapter 4. Parabolic triples — 19
  1. Definitions and basic facts — 19
  2. Parabolic Higgs bundles and parabolic triples — 20
  3. Extensions and deformations of parabolic triples — 21

Chapter 5. Critical values and flips — 25
  1. Critical values — 25
  2. Crossing critical values and universal extensions — 26
  3. Flips — 28

Chapter 6. Parabolic triples with $r_1 = 2$ and $r_2 = 1$ — 33
  1. Flips — 33
  2. Poincaré polynomial of moduli of triples — 35

Chapter 7. Critical submanifolds of type $(1,1,1)$ — 39
  1. Description of the critical submanifolds — 39
  2. The sum for fixed ϖ. — 42
  3. The sum over ϖ. — 44

Chapter 8. Critical submanifolds of type $(1,2)$ — 47
  1. Description of the critical submanifolds — 47
  2. The sum for fixed (ϖ, ε). — 48
  3. The sum over ϖ and ε. — 49

Chapter 9. Critical submanifolds of type $(2,1)$ — 51
  1. Description of the critical submanifolds — 51
  2. The sum for fixed (ϖ, ε). — 53
  3. The sum over ϖ and ε. — 53

Chapter 10. Betti numbers of the moduli space of rank three parabolic bundles   57
1. Notation   57
2. Holla's formula   58
3. The rank 3 case   58

Chapter 11. Betti numbers of the moduli space of rank three parabolic Higgs bundles   63
1. Poincaré polynomial   63
2. Special low genus cases   64

Chapter 12. The fixed determinant case   67
1. Preliminaries   67
2. Morse indices   68
3. Critical submanifolds of type $(1,1,1)$ 69
4. Parabolic triples of fixed determinant   72
5. Critical submanifolds of type $(1,2)$ and $(2,1)$ 75
6. Critical submanifolds of type (3) 76
7. Betti numbers of the fixed determinant moduli space   76

Bibliography   79

Abstract

Parabolic Higgs bundles on a Riemann surface are of interest for many reasons, one of them being their importance in the study of representations of the fundamental group of the punctured surface in the complex general linear group. In this paper we calculate the Betti numbers of the moduli space of rank 3 parabolic Higgs bundles with fixed and non-fixed determinant, using Morse theory. A key point is that certain critical submanifolds of the Morse function can be identified with moduli spaces of parabolic triples. These moduli spaces come in families depending on a real parameter and we carry out a careful analysis of them by studying their variation with this parameter. Thus we obtain in particular information about the topology of the moduli spaces of parabolic triples for the value of the parameter relevant to the study of parabolic Higgs bundles. The remaining critical submanifolds are also described: one of them is the moduli space of parabolic bundles, while the remaining ones have a description in terms of symmetric products of the Riemann surface. As another consequence of our Morse theoretic analysis, we obtain a proof of the parabolic version of a theorem of Laumon, which states that the nilpotent cone (the preimage of zero under the Hitchin map) is a Lagrangian subvariety of the moduli space of parabolic Higgs bundles.

Received by the editor January 7, 2005, and in revised form May 27, 2005.
2000 *Mathematics Subject Classification.* 14D20, 14H60.
Key words and phrases. Parabolic bundles, Higgs bundles, moduli spaces.
Partially supported by Ministerio de Educación y Tecnología and Consejo de Reitores das Universidades Portuguesas through Acción Integrada Hispano-Lusa HP-2000-0015 (Spain) / E–30/03 (Portugal) and by the European Research Training Networks EDGE (Contract no. HPRN-CT-2000-00101) and EAGER (Contract no. HPRN-CT-2000-00099). Members of VBAC (Vector Bundles on Algebraic Curves). Second author partially supported by Centro de Matemática da Universidade do Porto, financed by FCT (Portugal) through the programmes POCTI and POSI of the QCA III (2000–2006) with European Community (FEDER) and national funds. First and third authors partially supported by Ministerio de Educación y Ciencia (Spain) through Project BFM2000-0024.

CHAPTER 1

Introduction

Let X be a connected, smooth projective complex algebraic curve of genus g and let $D = p_1 + p_2 + \cdots + p_n$ be a divisor, with distinct points $p_1, \ldots, p_n$. Let K be the canonical bundle of X. A *parabolic Higgs bundle* is a pair (E, Φ), where E is a parabolic bundle, that is a holomorphic bundle over X together with a weighted flag in the fibre of E over each $p \in D$, and $\Phi : E \to E \otimes K(D)$ is a strongly parabolic homomorphism. This means that Φ is a meromorphic endomorphism valued one-form with simple poles along D whose residue at p is nilpotent with respect to the flag.

Like in the non-parabolic case, there is a stability criterion allowing the construction of moduli spaces of semistable parabolic Higgs bundles [39]. For generic weights, semistability and stability coincide and the moduli space is a smooth quasiprojective algebraic manifold. The goal of this paper is to compute the Betti numbers of this moduli space in the case in which the rank of the bundle is 3. The computation for rank 2 was carried out by Boden and Yokogawa [6], and Nasatyr and Steer [33] in the case of rational weights. In the non-parabolic case the Betti numbers had been previously computed by Hitchin [23] in rank 2 and Gothen [16] in rank 3. We have been informed by T. Hausel of a conjecture for the Betti numbers of the moduli space of parabolic Higgs bundles of any rank and degree, analogous to his conjecture for the case of non-parabolic Higgs bundles [20]. His formula gives the same result as ours in the cases that we have checked (cf. Remark 11.3), thus providing support for his conjecture.

Similarly to the non-parabolic case, the moduli space of parabolic Higgs bundles has an extremely rich geometric structure. It can be identified with the moduli space of solutions to the parabolic version of Hitchin's equations:

$$F(A)^\perp + [\Phi, \Phi^*] = 0 \quad \text{and} \quad \bar{\partial}_A \Phi = 0,$$

where A is a singular connection unitary with respect to a singular hermitian metric on E adapted to the parabolic structure (see Chapter 2 for details). The moduli space of parabolic Higgs bundles contains the total space of the cotangent bundle of the moduli space of parabolic bundles, whose natural holomorphic symplectic form can be extended to the whole moduli space. This form can be combined with the real symplectic form coming from the gauge-theoretic interpretation to endow the moduli space with a hyperkähler structure [26, 33].

An important motivation to study ordinary Higgs bundles comes from their relation with complex representations of the fundamental group of the curve. This is established by identifying the moduli space of solutions to Hitchin's equations with the moduli space of Higgs bundles [23, 36] as well as with the moduli space of complex connections with constant central curvature [11, 12]. In the parabolic case, there is a similar correspondence proved by Simpson [35]. This involves meromorphic complex connections with simple poles at the points and parabolic weights. At the topological side one has to consider filtered local systems. The natural context for the correspondence is a class larger than parabolic Higgs bundles in which the Higgs field Φ is allowed to be parabolic and not necessarily strongly parabolic. In other words, at a parabolic point the residue of Φ is parabolic with respect to the flag. Under this correspondence, parabolic Higgs bundles (those for which the Higgs field is strongly parabolic) are in bijection with meromorphic flat connections whose holonomy around each parabolic point defines a conjugacy class of an element in the unitary group. These, in turn, correspond to representations of the fundamental group of the punctured surface in the general linear group, which send a small loop around each parabolic point to an element conjugate to a unitary element.

The main tool for our computation of the Betti numbers, as in the previously studied cases, is the use of the Morse-theoretic techniques introduced by Hitchin [23]: the L^2-norm of the Higgs field defines a perfect Bott–Morse function on the moduli space. We have to compute the Poincaré series and the indices of the various critical subvarieties. In fact the calculation of the indices can be carried out for any rank, whereas the calculation of the Poincaré series of the critical subvarieties depends crucially on the rank 3 assumption. Here is a description of the paper.

In Chapter 2 we review the basic definitions and basic facts of parabolic Higgs bundles. In Chapter 3 we consider the Bott–Morse function on the moduli space and identify the critical subvarieties. These coincide with the fixed subvarieties under the action of S^1 on the moduli space given by multiplying the Higgs field. These in turn correspond, as shown by Simpson [35], to variations of Hodge structures, in particular the bundle has to be a direct sum of subbundles. We then compute the indices — this can be done for any rank and leads to a parabolic version of the theorem of Laumon, that the nilpotent cone in the moduli space of parabolic Higgs bundles is a Lagrangian subvariety. In the rank 3 case, the possible decompositions of the vector bundle in a sum of subbundles are of two types: a sum of three line bundles or a sum of a line bundle and a rank 2 vector bundle. The latter case gives rise to so-called parabolic triples. These have been introduced in [4] and generalise the triples studied in [8] and [9]. Through Chapters 4, 5 and 6 we study the moduli spaces of parabolic triples. They depend on a real parameter, relating to parabolic Higgs bundles when the value of this parameter is $2g-2$. To compute the Betti numbers for a given value of the parameter (in particular for $2g-2$) we follow the strategy introduced by Thaddeus in [37]. After characterising the moduli space for the largest value of the parameter, we need to analyse the changes when we cross a finite number of values until we get to the one we want. In Chapter 7 we compute the Poincaré polynomial and indices for the critical subvarieties for which the vector bundle is a sum of three line bundles. In Chapters 8 and 9 we do the other cases using the previous computations for the moduli space of parabolic triples. An important technical point is that the Betti numbers of the moduli space of parabolic Higgs bundles do not depend on the degree and the weights (certainly if the weights are generic). So we can choose the degree coprime with the rank and the weights as small as convenient, to facilitate our computations. In Chapter 10, based on the computations by Nitsure [34] and Holla [25] of the Betti numbers of the moduli space of parabolic bundles, we work out the formula for the rank 3 case. In Chapter 11 we collect all the computations, to give the Poincaré polynomial of the rank 3 moduli space of parabolic Higgs bundles. Finally, in Chapter 12, we compute the Poincaré polynomial of the rank 3 moduli space of parabolic Higgs bundles with fixed determinant. It is interesting to observe that, like in the non-parabolic case, and in contrast to the case of stable parabolic bundles, the Poincaré polynomial of the non-fixed determinant moduli space does not split as the product of those of the Jacobian and the fixed determinant moduli space. In particular, it follows that tensoring by a line bundle gives a non-trivial action of the group of elements of order three in the Jacobian on the cohomology of the fixed determinant moduli space with rational coefficients; in fact our methods allow us to determine precisely the non-invariant part of the rational cohomology.

Acknowledgements. We wish to thank Tamás Hausel and Nigel Hitchin for very useful comments.

CHAPTER 2

Parabolic Higgs bundles

1. Definitions and basic facts

Let X be a connected, smooth projective complex algebraic curve of genus g together with a finite (non-zero) number of *marked* distinct points $p_1, \ldots, p_n$. We will denote the divisor $D = p_1 + p_2 + \cdots + p_n$.

Let E be a holomorphic bundle over X. A parabolic structure on E consists of weighted flags

$$E_p = E_{p,1} \supset E_{p,2} \supset \cdots \supset E_{p,s_p} \supset E_{p,s_p+1} = 0,$$

$$0 \leq \alpha_1(p) < \cdots < \alpha_{s_p}(p) < 1,$$

over each $p \in D$. A holomorphic map $\phi : E \to F$ between parabolic bundles is called parabolic if $\alpha_i^E(p) > \alpha_j^F(p)$ (the superindex indicating to which parabolic bundle the weight corresponds) implies $\phi(E_{p,i}) \subset F_{p,j+1}$ for all $p \in D$. We call ϕ strongly parabolic if $\alpha_i^E(p) \geq \alpha_j^F(p)$ implies $\phi(E_{p,i}) \subset F_{p,j+1}$ for all $p \in D$.

We will abuse notation by simply writing E for a bundle with a parabolic structure.

This notion can be generalised to higher dimensions. In this case, a parabolic bundle consists of a holomorphic vector bundle together with a weighted holomorphic filtration of the restriction of the bundle to a fixed divisor. This generalisation is also relevant for us in the particular case in which the manifold is the product of the curve X and a higher dimensional manifold Y, and the divisor is $D \times Y$. Parabolic and strongly parabolic homomorphisms are defined in a similar way.

Also $\mathrm{ParHom}(E, F)$ and $\mathrm{SParHom}(E, F)$ will denote respectively the sheaves of parabolic and strongly parabolic homomorphisms from E to F.

Let $m_i(p) = \dim E_{p,i}/E_{p,i+1}$ be the multiplicity of $\alpha_i(p)$. It will sometimes be convenient to repeat each weight according to its multiplicity, i.e., we set $\tilde{\alpha}_1(p) = \ldots = \tilde{\alpha}_{m_1(p)}(p) = \alpha_1(p)$, etc. We then have weights $0 \leq \tilde{\alpha}_1(p) \leq \cdots \leq \tilde{\alpha}_r(p) < 1$, where $r = \mathrm{rk}(E)$ is the rank of E. Define the parabolic degree and parabolic slope of E by

$$\mathrm{pardeg}(E) = \deg(E) + \sum_{p \in D} \sum_{i=1}^{s_p} m_i(p)\alpha_i(p) = \deg(E) + \sum_{p \in D} \sum_{i=1}^{r} \tilde{\alpha}_i(p),$$

$$\mathrm{par}\mu(E) = \frac{\mathrm{pardeg}(E)}{\mathrm{rk}(E)}.$$

If F is a subbundle of E, then F inherits a parabolic structure by setting $F_{p,i} = F_p \cap E_{p,i}$ and discarding those weights of multiplicity zero. We call this the induced parabolic structure on F. In a similar manner, one can give a parabolic structure to the quotient E/F.

A parabolic bundle E is said to be stable if $\text{par}\mu(F) < \text{par}\mu(E)$ for all proper parabolic subbundles $F \subset E$. Semistability is defined by replacing the strict inequality by the weak inequality. For generic weights, stability and semistability are equivalent.

For parabolic bundles E there is a well-defined notion of dual E^*. This is done by considering the bundle $\text{Hom}(E, \mathcal{O}(-D))$, and at each $p \in D$, defining the filtration
$$E_p^* = E_{p,1}^* \supset \cdots \supset E_{p,s_p}^* \supset 0,$$
with $E_{p,i}^* = \text{Hom}(E_p/E_{p,s_p+2-i}, \mathcal{O}(-D)_p)$ and weights $1 - \alpha_s(p) < \cdots < 1 - \alpha_1(p)$. It is easy to prove that $E^{**} = E$, and $\text{pardeg}(E^*) = -\text{pardeg}(E)$.

There is also a notion of tensor product $\otimes^P$ of two parabolic bundles [**40**], which is best understood in terms of $\mathbb{R}$-filtered sheaves. Here we shall only use the case of tensoring a parabolic bundle E with a parabolic line bundle L. Let $\alpha_i(p)$ be the weights of E and $\beta(p)$ be the weights of L. Then the parabolic bundle $F = E \otimes^P L$ is, as a bundle, the kernel of
$$E \otimes L(D) \twoheadrightarrow \oplus_{p \in D} \left((E_p/E_{p,i_p}) \otimes L(D)_p \right),$$
where $i_p = \min\{s_p + 1, i \mid \alpha_i(p) + \beta(p) \geq 1\}$, $p \in D$. The weights of F are
(2.1)
$$\alpha_{i_p}(p) + \beta(p) - 1 < \cdots < \alpha_{s_p}(p) + \beta(p) - 1 < \alpha_1(p) + \beta(p) < \cdots < \alpha_{i_p-1}(p) + \beta(p),$$
with multiplicities $m_{i_p}(p), \ldots, m_{s_p}(p), m_1(p), \ldots, m_{i_p-1}(p)$. It is then easy to see that $\text{par}\mu(E \otimes^P L) = \text{par}\mu(E) + \text{par}\mu(L)$, whereas
$$\deg(E \otimes^P L) = \deg(E) + \text{rk}(E)\deg(L) + \sum_p \dim E_{p,i_p}.$$

We denote by K the canonical bundle on X. A *parabolic Higgs bundle* is a pair (E, Φ), where E is a parabolic bundle and $\Phi \in H^0(\text{SParEnd}(E) \otimes K(D))$, i.e. Φ is a meromorphic endomorphism valued one-form with simple poles along D whose residue at p is nilpotent with respect to the flag. We shall sometimes denote a parabolic Higgs bundle by $\mathbf{E} = (E, \Phi)$.

The notion of stability is extended to parabolic Higgs bundles in the usual way:
$$\text{par}\mu(F) < \text{par}\mu(E)$$
for all proper parabolic subbundles $F \subset E$ which are preserved by Φ. Semistability is defined by replacing the strict inequality by the weak inequality.

The standard properties of stable bundles also apply to parabolic Higgs bundles; for example, if $\mathbf{E}$ and $\mathbf{F}$ are stable parabolic Higgs bundles of the same parabolic slope, then there are no parabolic maps between them unless they are isomorphic, and the only parabolic endomorphisms of a stable parabolic Higgs bundle are the scalar multiples of the identity.

We shall say that the weights are *generic* when every semistable parabolic Higgs bundle is automatically stable, i.e., when there are no properly semistable Higgs bundles. Let us fix (generic) weights $\alpha_i(p)$ and topological invariants $\text{rk}(E)$ and $\deg(E)$. The moduli space $\mathcal{M}$ of stable parabolic Higgs bundles was constructed using Geometric Invariant Theory by Yokogawa [**39**, **40**], who also showed that it is a smooth irreducible complex variety. The moduli space $\mathcal{M}$ contains the cotangent bundle of the moduli space of stable parabolic bundles.

The following result will facilitate the computation of the Betti numbers of $\mathcal{M}$.

PROPOSITION 2.1. *Fix the rank r. For different choices of degrees and generic weights, the moduli spaces of parabolic Higgs bundles have the same Betti numbers.*

PROOF. For fixed degree, it is a consequence of the results of Thaddeus [38] that the moduli spaces for different generic weights have the same Betti numbers, as we now explain. The space of weights is divided into chambers by a finite number of hyperplanes, or *walls*, and in each chamber the moduli spaces are isomorphic. Call the moduli spaces on each side of a wall $\mathcal{M}^+$ and $\mathcal{M}^-$, respectively (here, and in the following, we use the notation of [38]). Thaddeus proves that $\mathcal{M}^+$ and $\mathcal{M}^-$ have a common blow-up with the same exceptional divisor. The loci in $\mathcal{M}^\pm$ to be blown up (*flip loci*, in our language) are isomorphic to projective bundles $\mathbb{P}U^\pm$ over a product $\mathcal{N}^+ \times \mathcal{N}^-$ of moduli spaces of lower rank parabolic Higgs bundles. This is similar[1] to the situation in [37] (for ordinary triples) and our analysis in Chapter 5 below (for parabolic triples), and shows that the difference between the Poincaré polynomials of $\mathcal{M}^+$ and $\mathcal{M}^-$ equals the difference between the Poincaré polynomials of the respective flip loci (cf. [15, p. 605]). However, in the case of parabolic Higgs bundles something special happens, namely the bundles U^+ and U^- are dual to each other (see the paragraph of [38] preceding (5.6)). Hence $\mathbb{P}U^+$ and $\mathbb{P}U^-$ are projective bundles of the same rank, over the same base. But the Poincaré polynomial of a projective bundle splits as the product of the Poincaré polynomial of the base and the Poincaré polynomial of projective space. Thus the flip loci in $\mathcal{M}^+$ and $\mathcal{M}^-$ have the same Betti numbers and, therefore, the moduli spaces $\mathcal{M}^+$ and $\mathcal{M}^-$ themselves have the same Betti numbers.

To extend the result to moduli spaces of parabolic Higgs bundles with different degrees, we proceed as follows. Fix any parabolic line bundle L with degree d_L and weights $\beta(p)$. Then the map

$$(E, \Phi) \mapsto (E \otimes^P L, \Phi)$$

gives an *isomorphism* between the moduli space of parabolic Higgs bundles of rank r, degree Δ and weights $\alpha_i(p)$ and the moduli space of parabolic Higgs bundles of rank r, degree $\Delta + r d_L + \sum_p \sum_{i \geq i_p} m_i(p)$, and weights given by (2.1). Choosing weights of multiplicity one and a suitable parabolic line bundle L we see that moduli spaces of parabolic Higgs bundles for different degrees are isomorphic. Since we already know that the Betti numbers are independent of the (generic) weights for fixed degree, this concludes the proof. □

2. Deformation theory

The deformation theory of parabolic Higgs bundles was worked out by Yokogawa [40]; see also Thaddeus [38] and Biswas and Ramanan [5]. Everything in this section is essentially contained in these references but we shall find it convenient to give an exposition tailored to our purposes. Let $\mathbf{E} = (E, \Phi)$ and $\mathbf{F} = (F, \Psi)$ be parabolic Higgs bundles. We define a complex of sheaves

$$C^\bullet(\mathbf{E}, \mathbf{F}): \quad \operatorname{ParHom}(E, F) \to \operatorname{SParHom}(E, F) \otimes K(D)$$
$$f \mapsto (f \otimes 1)\Phi - \Psi f,$$

and write $C^\bullet(\mathbf{E}) = C^\bullet(\mathbf{E}, \mathbf{E})$.

---

[1]In these cases it is the stability parameter which is varied, rather than the parabolic weights.

PROPOSITION 2.2.   (i) *The space of infinitesimal deformations of a parabolic Higgs bundle* **E** *is naturally isomorphic to the first hypercohomology group of the complex*

$$C^\bullet(\mathbf{E}): \quad \operatorname{ParEnd}(E) \xrightarrow{[-,\Phi]} \operatorname{SParEnd}(E) \otimes K(D)$$
$$f \longmapsto (f \otimes 1)\Phi - \Phi f.$$

Thus the tangent space to $\mathcal{M}$ *at a point represented by a stable parabolic Higgs bundle* **E** *is isomorphic to* $\mathbb{H}^1(C^\bullet(\mathbf{E}))$.

(ii) *The space of homomorphisms between parabolic Higgs bundles* **E** *and* **F** *is naturally isomorphic to the zeroth hypercohomology group* $\mathbb{H}^0(C^\bullet(\mathbf{E}, \mathbf{F}))$.

(iii) *The space of extensions* $0 \to \mathbf{E}' \to \mathbf{E} \to \mathbf{E}'' \to 0$ *of parabolic Higgs bundles* **E**' *and* **E**'' *is naturally isomorphic to the first hypercohomology* $\mathbb{H}^1(C^\bullet(\mathbf{E}'', \mathbf{E}'))$.

(iv) *There is a long exact sequence*

$$0 \to \mathbb{H}^0(C^\bullet(\mathbf{E}, \mathbf{F})) \to H^0(\operatorname{ParHom}(E, F)) \to H^0(\operatorname{SParHom}(E, F) \otimes K(D))$$
(2.2) $$\to \mathbb{H}^1(C^\bullet(\mathbf{E}, \mathbf{F})) \to H^1(\operatorname{ParHom}(E, F)) \to H^1(\operatorname{SParHom}(E, F) \otimes K(D))$$
$$\to \mathbb{H}^2(C^\bullet(\mathbf{E}, \mathbf{F})) \to 0.$$

PROOF. For proofs of (i) – (iii) see Thaddeus [38]. The proof of (iv) follows by a standard argument in Higgs bundle theory (see, e.g., Biswas and Ramanan [5]). □

As for ordinary Higgs bundles, duality plays an important role for parabolic Higgs bundles. The results of the following proposition are consequences of the theory developed by Yokogawa (cf. (3.1) and Proposition 3.7 of [40], see also § 3 of Thaddeus [38] and § 5 of Bottacin [7]).

PROPOSITION 2.3.   (i) *Let E and F be parabolic bundles. The sheaves* $\operatorname{ParHom}(E, F)$ *and* $\operatorname{SParHom}(F, E(D))$ *are naturally dual.*

(ii) *Let* **E** *and* **F** *be parabolic Higgs bundles. Then there is a natural isomorphism*

$$\mathbb{H}^i(C^\bullet(\mathbf{E}, \mathbf{F})) \cong \mathbb{H}^{2-i}(C^\bullet(\mathbf{F}, \mathbf{E}))^*.$$

In particular we obtain a natural isomorphism $T_\mathbf{E}\mathcal{M} \cong T^*_\mathbf{E}\mathcal{M}$ *for a stable parabolic Higgs bundle* **E**.

PROOF. We just show how (i) implies (ii). From (i) it follows that the dual complex $C^\bullet(\mathbf{E}, \mathbf{F})^*$ is related to the original one by

$$C^\bullet(\mathbf{E}, \mathbf{F})^* \otimes K \cong C^\bullet(\mathbf{F}, \mathbf{E}).$$

Thus, Serre duality for hypercohomology and Proposition 2.2 (i) give statement (ii) of the present proposition. □

Next we shall show how these results can be used to calculate the dimension of the moduli space of parabolic Higgs bundles. The result is well known but it seems worthwhile to include the calculation here, since we shall use the same ideas again below. First we introduce some convenient notation for parabolic bundles E and F as follows. We denote by $P_p(E, F)$ the subspace of $\operatorname{Hom}(E_p, F_p)$ consisting of parabolic maps and by $N_p(E, F)$ the subspace of strictly parabolic maps. We also write $P_D(E, F) = \bigoplus P_p(E, F)$ and $N_D(E, F) = \bigoplus N_p(E, F)$. When there is no

risk of confusion we shall omit the parabolic bundles E and F from the notation. We then have short exact sequences of sheaves
$$0 \to \mathrm{ParHom}(E,F) \to \mathrm{Hom}(E,F) \to \mathrm{Hom}(E_D, F_D)/P_D(E,F) \to 0,$$
and
$$0 \to \mathrm{SParHom}(E,F) \to \mathrm{Hom}(E,F) \to \mathrm{Hom}(E_D, F_D)/N_D(E,F) \to 0.$$
Thus we can calculate the Euler characteristics of the sheaves $\mathrm{ParHom}(E,F)$ and $\mathrm{SParHom}(E,F)$ as follows:

(2.3)
$$\chi(\mathrm{ParHom}(E,F)) = \chi(\mathrm{Hom}(E,F)) + \sum_{p \in D} (\dim P_p - \mathrm{rk}(E)\,\mathrm{rk}(F)),$$
$$\chi(\mathrm{SParHom}(E,F)) = \chi(\mathrm{Hom}(E,F)) + \sum_{p \in D} (\dim N_p - \mathrm{rk}(E)\,\mathrm{rk}(F)).$$

With these preliminaries in place we can calculate the dimension of the moduli space.

PROPOSITION 2.4. *The complex dimension of the moduli space $\mathcal{M}$ of stable rank r parabolic Higgs bundles is*
$$r^2(2g-2) + 2 + 2\sum_p f_p,$$
where $r = \mathrm{rk}(E)$ and $f_p = \tfrac{1}{2}\bigl(r^2 - \sum_i m_i(p)^2\bigr)$.

PROOF. Since $\mathbf{E}$ is stable, its only endomorphisms are the scalars. Hence, using Proposition 2.2 (ii) and the duality statement Proposition 2.3 (ii), we have that $\dim \mathbb{H}^0(C^\bullet(\mathbf{E})) = \dim \mathbb{H}^2(C^\bullet(\mathbf{E})) = 1$. It follows that the dimension of the moduli space is
$$\begin{aligned}
\dim \mathcal{M} &= \dim \mathbb{H}^1(C^\bullet(\mathbf{E})) \\
&= 2 - \chi(C^\bullet(\mathbf{E})) \\
&= 2 - \chi(\mathrm{ParEnd}(E)) + \chi(\mathrm{SParEnd}(E) \otimes K(D)),
\end{aligned}$$
where in the last equality we have used the long exact sequence (2.2). From this we obtain the result by using equations (2.3), the fact that $\dim P_p - \dim N_p = \sum_i m_i(p)^2$ and the Riemann–Roch formula. □

3. Parabolic Higgs bundles and gauge theory

Our main goal is to study the topology of $\mathcal{M}$. To do this we need the gauge-theoretic interpretation of this moduli space in terms of solutions to Hitchin's equations due to Simpson [**35**]. The construction of the moduli space from this point of view is due to Konno [**26**]. Let E be a smooth parabolic vector bundle of rank r and fix a hermitian metric h in E which is smooth in $X \smallsetminus D$ and whose (degenerate) behaviour around the punctures is given as follows. We say that a local frame $\{e_1, \dots, e_r\}$ for E around p respects the flag at p if $E_{p,i}$ is spanned by the vectors $\{e_{M_i+1}(p), \dots, e_r(p)\}$, where $M_i = \sum_{j \leq i} m_j$. Let z be a local coordinate around p such that $z(p) = 0$. We require that h be of the form
$$h = \begin{pmatrix} |z|^{2\tilde{\alpha}_1} & & 0 \\ & \ddots & \\ 0 & & |z|^{2\tilde{\alpha}_r} \end{pmatrix}$$

with respect to some local frame around p which respects the flag at p. We denote the space of smooth $\bar{\partial}$-operators on E by $\mathscr{C}$ and the space of associated h-unitary connections by $\mathscr{A}$. Note that the unitary connection associated to a smooth $\bar{\partial}_A$ via the hermitian metric h is singular around the punctures: if we write $z = \rho \exp(\mathbf{i}\theta)$ and $\{e_i\}$ is the local frame used in the definition of h, then with respect to the local frame $\{\epsilon_i = e_i/|z|^{\tilde{\alpha}_i}\}$, the connection is of the form

$$d_A = d + \mathbf{i} \begin{pmatrix} \tilde{\alpha}_1 & & 0 \\ & \ddots & \\ 0 & & \tilde{\alpha}_r \end{pmatrix} d\theta + A',$$

where A' is regular.

We denote the space of Higgs fields by $\mathbf{\Omega} = \Omega^{1,0}(\mathrm{SParEnd}(E) \otimes \mathcal{O}(D))$, the group of complex parabolic gauge transformations by $\mathscr{G}_{\mathbb{C}}$ and the subgroup of h-unitary parabolic gauge transformations by $\mathscr{G}$.

Following Biquard [3], Konno introduces certain weighted Sobolev norms; we denote the corresponding Sobolev completions of the spaces defined above by $\mathscr{C}_1^p$, $\mathbf{\Omega}_1^p$, $(\mathscr{G}_{\mathbb{C}})_2^p$ and $\mathscr{G}_2^p$ (the detailed definitions are not important to us so we refer to [26] for them). Let

$$\mathscr{H} = \{(\bar{\partial}_A, \Phi) \in \mathscr{C} \times \mathbf{\Omega} \mid \bar{\partial}_A \Phi = 0\}$$

and let $\mathscr{H}_1^p$ be the corresponding subspace of $\mathscr{C}_1^p \times \mathbf{\Omega}_1^p$. Then $\mathscr{H}_1^p$ carries a hyper-Kähler metric induced by h. Let $F(A)^{\perp}$ denote the trace-free part of the curvature of the h-unitary connection A corresponding to $\bar{\partial}_A$ and let Φ^* be the adjoint with respect to h. One can then consider the moduli space $\mathcal{S}$ defined by the subspace of $\mathscr{H}_1^p$ satisfying *Hitchin's equation* (modulo $\mathscr{G}_2^p$),

$$\mathcal{S} = \{(\bar{\partial}_A, \Phi) \in \mathscr{H}_1^p \mid F(A)^{\perp} + [\Phi, \Phi^*] = 0\}/\mathscr{G}_2^p,$$

where the equation is only defined on $X \setminus D$. Konno proves that $\mathcal{S}$ is a hyper-Kähler quotient and that it can be naturally identified with the moduli space

$$\mathcal{M} = \mathscr{H}_1^p/(\mathscr{G}_{\mathbb{C}})_2^p.$$

Furthermore, Konno proves that the natural map $\mathscr{H}/\mathscr{G}_{\mathbb{C}} \to \mathscr{H}_1^p/(\mathscr{G}_{\mathbb{C}})_2^p$ is a diffeomorphism.

CHAPTER 3

Morse theory on the moduli space

1. The Morse function

The non-zero complex numbers $\mathbb{C}^*$ act on the moduli space $\mathcal{M}$ via the map $\lambda \cdot (E, \Phi) = (E, \lambda\Phi)$. However, to have an action on the set of solutions to Hitchin's equations, one must restrict to the action of $S^1 \subset \mathbb{C}^*$. Obviously the identification $\mathcal{S} \cong \mathcal{M}$ respects the circle action and thus we have a circle action on this hyper-Kähler manifold. With respect to one of the complex structures (coinciding with the one on $\mathcal{M}$) this is a Hamiltonian action and the associated moment map is

$$[(A, \Phi)] \mapsto -\tfrac{1}{2}\|\Phi\|^2 = -\mathbf{i}\int_X \mathrm{Tr}\,(\Phi\Phi^*). \tag{3.1}$$

We shall, however, prefer to consider the positive function

$$f([A, \Phi]) = \tfrac{1}{2}\|\Phi\|^2. \tag{3.2}$$

In the case of non-parabolic Higgs bundles, Hitchin [**23**] proved that this is a proper map, using Uhlenbeck's compactness theorem. It was observed by Boden and Yokogawa [**6**] that the same argument works in the parabolic case, by using the parabolic analogue of Uhlenbeck's theorem, proved by Biquard [**3**]. Thus we have the following result.

PROPOSITION 3.1. *The map $f\colon \mathcal{M} \to \mathbb{R}$ is proper.* □

Next we recall a general result of Frankel [**13**], which was first used in the context of moduli spaces of Higgs bundles by Hitchin [**23**].

THEOREM 3.2. *Let $\tilde{f}\colon M \to \mathbb{R}$ be a proper moment map for a Hamiltonian circle action on a Kähler manifold M. Then $\tilde{f}$ is a perfect Bott–Morse function.* □

The following result on the Morse indices of such a Morse function is implicit in Frankel's paper.

PROPOSITION 3.3. *In the situation of Theorem 3.2, the critical points of $\tilde{f}$ are exactly the fixed points of the circle action. Moreover, the eigenvalue l subspace for the Hessian of $\tilde{f}$ is the same as the weight $-l$ subspace for the infinitesimal circle action on the tangent space. In particular, the Morse index of $\tilde{f}$ at a critical point equals the dimension of the positive weight space of the circle action on the tangent space.*

PROOF. The condition for $\tilde{f}$ to be a moment map is that

$$\mathrm{grad}(\tilde{f}) = IX,$$

where X is the vector field generating the circle action and I is the complex structure on M. Hence p is a critical point of $\tilde{f}$ if and only if it is fixed under the circle action.

Let ∇ be the Levi-Civita connection on M, then the Hessian $H_{\tilde{f}}$ of $\tilde{f}$ at p is the quadratic form associated to the symmetric endomorphism $\nabla(\text{grad}(\tilde{f}))_p$ of T_pM. Let $Y_p \in T_pM$ and let the vector field Y be an extension of Y_p around p. Then we have

$$H_{\tilde{f}}(Y_p) = \nabla_{Y_p}(IX)$$
$$= \nabla_{IX_p}(Y) - [IX, Y]_p$$
$$= -[IX, Y]_p,$$

where we have used that $X_p = 0$. On the other hand it is easy to see (cf. [13]) that the infinitesimal circle action on T_pM is given by $Y_p \mapsto [Y, X]_p$. It follows that the eigenvalues of $H_{\tilde{f}}$ are exactly minus the weights of the circle action on T_pM. □

Thus we must identify the fixed point set of the action of $S^1 \subset \mathbb{C}^*$ on $\mathcal{M}$. This was done by Simpson and is analogous to what happens for ordinary Higgs bundles.

2. Fixed points of the S^1 action on the moduli space

PROPOSITION 3.4 ([35, Theorem 8]). *The equivalence class of a stable parabolic Higgs bundle (E, Φ) is fixed under the action of S^1 if and only if it is a parabolic complex variation of Hodge structure. This means that E has a direct sum decomposition*

$$E = E_0 \oplus \cdots \oplus E_m$$

as parabolic bundles, such that Φ is strongly parabolic and of degree one with respect to this decomposition, in other words the restriction $\Phi_l = \Phi_{|E_l}$ belongs to

$$H^0(\text{SParHom}(E_l, E_{l+1}) \otimes K(D)).$$

Furthermore, stability implies that $\Phi_l \neq 0$ for $l = 0, \ldots, m-1$. The type of the parabolic complex variation of Hodge structure is the vector $(\text{rk}(E_0), \ldots, \text{rk}(E_m))$.
□

REMARK 3.5. If $m = 0$, then $E = E_0$ and $\Phi = 0$, corresponding to the obvious fixed points $(E, 0)$, with E a stable parabolic bundle.

The following important fact was also noted by Simpson. For a proof see [1, Proposition 3.11] (in fact, this deals with the ordinary case but the argument can easily be adapted to the parabolic case).

PROPOSITION 3.6. *A parabolic complex variation of Hodge structure $(E = \bigoplus E_l, \Phi)$ is stable as a parabolic Higgs bundle if and only if the stability condition is satisfied for subbundles of E which respect the decomposition $E = \bigoplus E_l$.* □

Next we need to calculate the weights of the circle action on the tangent space to $\mathcal{M}$ at a critical point of f, represented by $\mathbf{E} = (\bigoplus E_l, \Phi)$. By the characterization of the critical points provided by Propositions 3.3 and 3.4, we have decompositions

$$\text{ParEnd}(E) = \bigoplus_{l=-m}^{m} U_l, \qquad \text{SParEnd}(E) = \bigoplus_{l=-m}^{m} \hat{U}_l,$$

where we use the notation

$$U_l = \bigoplus_{j-i=l} \text{ParHom}(E_i, E_j), \qquad \hat{U}_l = \bigoplus_{j-i=l} \text{SParHom}(E_i, E_j).$$

2. FIXED POINTS OF THE S^1 ACTION ON THE MODULI SPACE

We get a corresponding decomposition of the deformation complex

$$C^\bullet(\mathbf{E}) = \bigoplus_{l=-m-1}^{m} C^\bullet(\mathbf{E})_l,$$

where $C^\bullet(\mathbf{E})_l$ denotes the subcomplex

$$C^\bullet(\mathbf{E})_l: \quad U_l \to \hat{U}_{l+1} \otimes K(D).$$

With this notation we have the following result.

PROPOSITION 3.7. *Let $\mathbf{E} = (\bigoplus E_l, \Phi)$ represent a fixed point of the circle action on $\mathcal{M}$. Then the weight l subspace of $T_\mathbf{E}\mathcal{M}$ is isomorphic to the first hypercohomology $\mathbb{H}^1(C^\bullet(\mathbf{E})_{-l})$.*

PROOF. It is clear that the derivative of the circle action at $\mathbf{E} = (E, \Phi)$ is induced by the following map of deformation complexes $C^\bullet(E, \Phi) \to C^\bullet(E, e^{i\theta}\Phi)$:

$$\begin{array}{ccccc}
C^\bullet(E, \Phi): & \bigoplus U_l & \xrightarrow{[-, \Phi]} & \bigoplus \hat{U}_{l+1} \otimes K(D) \\
& \downarrow e^{i\theta} & \downarrow 1 & \downarrow e^{i\theta} \\
C^\bullet(E, e^{i\theta}\Phi): & \bigoplus U_l & \xrightarrow{[-, e^{i\theta}\Phi]} & \bigoplus \hat{U}_{l+1} \otimes K(D).
\end{array}$$

In order to work out the circle action on $T_\mathbf{E}\mathcal{M}$ from this we need to determine the identification $\mathbb{H}^1(C^\bullet(E, \Phi)) \cong \mathbb{H}^1(C^\bullet(E, e^{i\theta}\Phi))$ induced by the isomorphism between (E, Φ) and $(E, e^{i\theta}\Phi)$. But it is easy to write down such an isomorphism f_θ: with respect to the decomposition $E = \bigoplus E_l$ we can define f_θ to be multiplication by $e^{il\theta}$ on E_l. The corresponding isomorphism between the complexes $C^\bullet(E, \Phi)$ and $C^\bullet(E, e^{i\theta}\Phi)$ is given by the adjoint $\mathrm{Ad}(f_\theta): \psi \mapsto f_\theta \psi f_\theta^{-1}$. Note that f_θ is unique up to multiplication by scalars and hence $\mathrm{Ad}(f_\theta)$ is unique. Since $\mathrm{Ad}(f_\theta)$ is multiplication by $e^{il\theta}$ on both U_l and $\hat{U}_l$, we can write down the induced isomorphism of complexes; the piece in degree l is given by

$$\begin{array}{ccccc}
C^\bullet(E, \Phi)_l: & U_l & \xrightarrow{[-, \Phi]} & \hat{U}_{l+1} \otimes K(D) \\
& \downarrow \mathrm{Ad}(f_\theta) & \downarrow e^{il\theta} & \downarrow e^{i(l+1)\theta} \\
C^\bullet(E, e^{i\theta}\Phi)_l: & U_l & \xrightarrow{[-, e^{i\theta}\Phi]} & \hat{U}_{l+1} \otimes K(D).
\end{array}$$

It follows that the derivative of the action of the map $e^{i\theta}$ is the endomorphism of $\mathbb{H}^1(C^\bullet(E, \Phi))$ induced by the composite map of complexes

$$C^\bullet(E, \Phi) \xrightarrow{e^{i\theta}} C^\bullet(E, e^{i\theta}\Phi) \xrightarrow{\mathrm{Ad}(f_\theta)^{-1}} C^\bullet(E, \Phi),$$

whose degree l piece is

$$\begin{array}{ccccc}
C^\bullet(E, \Phi)_l: & U_l & \xrightarrow{[-, \Phi]} & \hat{U}_{l+1} \otimes K(D) \\
& \downarrow & \downarrow e^{-il\theta} & \downarrow e^{-il\theta} \\
C^\bullet(E, \Phi)_l: & U_l & \xrightarrow{[-, \Phi]} & \hat{U}_{l+1} \otimes K(D).
\end{array}$$

Thus $\mathbb{H}^1(C^\bullet(E, \Phi)_l)$ is isomorphic to the weight $-l$ subspace of $\mathbb{H}^1(C^\bullet(E, \Phi)) \cong T_\mathbf{E}\mathcal{M}$. $\square$

Summarizing the results of this chapter so far, we obtain the following.

THEOREM 3.8. *The function* $f\colon \mathcal{M} \to \mathbb{R}$ *defined by* $f([A, \Phi]) = \frac{1}{2}\|\Phi\|^2$ *is a perfect Bott–Morse function. A parabolic Higgs bundle* (E, Φ) *represents a critical point of* f *if and only if it is a parabolic complex variation of Hodge structure, i.e.,* $E = \bigoplus_{l=0}^{m} E_l$ *with* $\Phi_l = \Phi_{|E_l}\colon E_l \to E_{l+1} \otimes K(D)$ *strongly parabolic (where* $\Phi = 0$ *if and only if* $m = 0$*). The tangent space to* $\mathcal{M}$ *at a critical point* $\mathbf{E}$ *decomposes as*
$$T_{\mathbf{E}}\mathcal{M} = \bigoplus_{l=-m}^{m+1} T_{\mathbf{E}}\mathcal{M}_l,$$
where the eigenvalue l *subspace of the Hessian of* f *is*
$$T_{\mathbf{E}}\mathcal{M}_l \cong \mathbb{H}^1(C^\bullet(E, \Phi)_{-l}).$$

PROOF. Immediate from Propositions 3.3 and 3.7. Note that since our Morse function f is *minus* the moment map $\tilde{f}$ (cf. (3.1) and (3.2)), the eigenvalue l subspace of the Hessian coincides with the weight l subspace for the circle action (with the *same* sign). □

3. Morse indices

PROPOSITION 3.9. (i) *There is a natural isomorphism*
$$\mathbb{H}^1(C^\bullet(\mathbf{E})_l) \cong \mathbb{H}^1(C^\bullet(\mathbf{E})_{-l-1})^*$$
and hence a natural isomorphism
$$T_{\mathbf{E}}\mathcal{M}_l \cong (T_{\mathbf{E}}\mathcal{M}_{1-l})^*.$$

(ii) *If* $\mathbf{E}$ *is stable, then we have*
$$\mathbb{H}^0(C^\bullet(\mathbf{E})_l) = \begin{cases} \mathbb{C} & \text{if } l = 0, \\ 0 & \text{otherwise,} \end{cases}$$
and
$$\mathbb{H}^2(C^\bullet(\mathbf{E})_l) = \begin{cases} \mathbb{C} & \text{if } l = -1, \\ 0 & \text{otherwise.} \end{cases}$$

PROOF. (i) It follows from Proposition 2.3 (i) that there is an isomorphism of complexes
$$(C^\bullet(\mathbf{E})_l)^* \otimes K \cong C^\bullet(\mathbf{E})_{-l-1}.$$
Hence Serre duality for hypercohomology gives the first isomorphism of the statement. The second isomorphism is now immediate from the last statement of Theorem 3.8.

(ii) When $\mathbf{E}$ is stable we have that $\mathbb{H}^0(C^\bullet(\mathbf{E})) \cong \mathbb{C}$, generated by the identity endomorphism of $\mathbf{E}$, and hence the first statement follows. For the same reason as in the proof of (i) we have the isomorphism $\mathbb{H}^0(C^\bullet(\mathbf{E})_l) \cong \mathbb{H}^2(C^\bullet(\mathbf{E})_{-l-1})^*$ and thus the second statement follows from the first. □

COROLLARY 3.10. *Let* $\mathbf{E}$ *represent a critical point of* f*, let* $T_{\mathbf{E}}\mathcal{M}_{\leq 0}$ *be the subspace of the tangent space on which the Hessian of* f *has eigenvalues less than or equal to zero and let* $T_{\mathbf{E}}\mathcal{M}_{>0}$ *be the subspace on which the Hessian of* f *has eigenvalues greater than zero. Then*
$$T_{\mathbf{E}}\mathcal{M}_{\leq 0} \cong (T_{\mathbf{E}}\mathcal{M}_{>0})^*$$

under the isomorphism of Proposition 2.3 (ii). It follows that the dimension of $T_{\mathbf{E}}\mathcal{M}_{\leq 0}$ is half the dimension of the moduli space, i.e.,

$$\dim T_{\mathbf{E}}\mathcal{M}_{\leq 0} = r^2(g-1) + 1 + \sum_p f_p.$$

PROOF. Immediate from Propositions 3.9 and 2.4. □

PROPOSITION 3.11. *Let the parabolic Higgs bundle* $\mathbf{E} = (E, \Phi)$ *represent a critical point of f. Then the Morse index of f at this point is*

$$\lambda_{\mathbf{E}} = r^2(2g-2) + 2\sum_p f_p + 2\chi(C^{\bullet}(\mathbf{E})_0)$$

$$= r^2(2g-2) + 2\sum_p f_p + 2\sum_{l=0}^{m} \chi\bigl(\mathrm{ParEnd}(E_l)\bigr)$$

$$- 2\sum_{l=0}^{m-1} \chi\bigl(\mathrm{SParHom}(E_l, E_{l+1}) \otimes K(D)\bigr)$$

$$= r^2(2g-2) + 2\sum_p f_p + 2\sum_{l=0}^{m}\Bigl[(1-g-n)\mathrm{rk}(E_l)^2 + \sum_p \dim P_p(E_l, E_l)\Bigr]$$

$$+ 2\sum_{l=0}^{m-1}\Bigl[(1-g)\mathrm{rk}(E_l)\mathrm{rk}(E_{l+1}) - \mathrm{rk}(E_l)\deg(E_{l+1}) + \mathrm{rk}(E_{l+1})\deg(E_l)$$

$$- \sum_p \dim N_p(E_l, E_{l+1})\Bigr],$$

where $E = \bigoplus_{l=0}^{m} E_l$ *with* $\Phi \in H^0(\mathrm{SParHom}(E_l, E_{l+1}) \otimes K(D))$.

PROOF. Since we are calculating real dimensions, the Morse index is twice the dimension of $T_{\mathbf{E}}\mathcal{M}_{<0}$, the subspace on which the Hessian of f has negative eigenvalues. Hence Corollary 3.10 shows that

$$\frac{1}{2}\lambda_{\mathbf{E}} = \dim T_{\mathbf{E}}\mathcal{M}_{<0}$$
$$= \dim T_{\mathbf{E}}\mathcal{M}_{\leq 0} - \dim T_{\mathbf{E}}\mathcal{M}_0$$
$$= r^2(g-1) + 1 + \sum_p f_p - \dim T_{\mathbf{E}}\mathcal{M}_0.$$

On the other hand from Proposition 3.9 (ii) we have that $\mathbb{H}^0(C^{\bullet}(\mathbf{E})_0) = \mathbb{C}$, while $\mathbb{H}^2(C^{\bullet}(\mathbf{E})_0) = 0$. Hence Theorem 3.8 shows that we have

$$\dim T_{\mathbf{E}}\mathcal{M}_0 = \dim \mathbb{H}^1(C^{\bullet}(\mathbf{E})_0)$$
$$= 1 - \chi(C^{\bullet}(\mathbf{E})_0),$$

and this finishes the proof of the first identity of the statement of the proposition. The rest can be deduced from the long exact sequence in hypercohomology for the complex $C^{\bullet}(\mathbf{E})_0$, analogous to (2.2), and using the same method as in the proof of Proposition 2.4. □

REMARK 3.12. Obviously, the absolute minima is for $m = 0$, for which the computation in Proposition 3.11 naturally gives

$$\lambda_{(E,0)} = r^2(2g-2) + 2\sum_p f_p + 2\chi(\operatorname{ParEnd}(E))$$
$$= r^2(2g-2) + 2\sum_p f_p + 2r^2(1-g) + 2\sum_p(r^2 - f_p - r^2)$$
$$= 0.$$

4. Rank three parabolic Higgs bundles

Now we turn our attention to the moduli space $\mathcal{M}$ of parabolic Higgs bundles of rank three. Let (E, Φ) be a critical point of f. By Theorem 3.8, the only possibilities that we have in this situation are:

(a) E is a stable rank three parabolic Higgs bundle and $\Phi = 0$.
(b) $E = E_0 \oplus E_1 \oplus E_2$ where E_l are parabolic line bundles. These are line bundles with weights at each $p \in D$. The map Φ decomposes as strongly parabolic maps $\Phi_0 : E_0 \to E_1 \otimes K(D)$ and $\Phi_1 : E_1 \to E_2 \otimes K(D)$.
(c) $E = E_0 \oplus E_1$ where E_0 is a parabolic line bundle L and E_1 is a rank 2 parabolic bundle. Here Φ gives a strongly parabolic map $\Phi_0 : L \to E_1 \otimes K(D)$.
(d) $E = E_0 \oplus E_1$ where E_0 is a rank 2 parabolic bundle and E_1 is a parabolic line bundle L. Here Φ gives a strongly parabolic map $\Phi_0 : E_0 \to L \otimes K(D)$.

In case (a) the corresponding critical subvariety can obviously be identified with the moduli space of ordinary parabolic bundles. Its Betti numbers can be computed from a formula given by Nitsure [**34**] and Holla [**25**]. In Chapter 10 we work out explicitly what their formula gives for the Poincaré polynomial in our situation of rank three parabolic bundles. Case (b) involves basically line bundles and divisors and can be dealt with easily [**29**]. The other two cases, (c) and (d), are more involved. They are particular cases of objects called *parabolic triples*, which have been introduced and studied from a gauge-theoretic point of view in [**4**], and will be studied in Chapter 4.

5. Laumon's Theorem for parabolic Higgs bundles

At this point we make a small digression in order to deduce, following Hausel, a parabolic version of a Theorem of Laumon from the analysis leading to our calculation of the Morse indices.

As in the non-parabolic case studied by Hitchin [**23, 24**], there is a *Hitchin map*

$$\chi : \mathcal{M} \to B = \bigoplus_{i=1}^r H^0(K(D)^i)$$

defined by taking the parabolic Higgs bundle (E, Φ) to the characteristic polynomial of Φ. Since the Higgs field is strictly parabolic, this map takes values in a subspace of B of dimension $r^2(g-1) + 1 + \sum_p f_p$, i.e., half the dimension of $\mathcal{M}$. The Hitchin map defines an algebraic completely integrable system. This means that the $r^2(g-1) + 1 + \sum_p f_p$ functions defined by χ Poisson commute, their differentials are linearly independent and the generic fibre of χ is an open set in an abelian variety. In fact $\mathcal{M}$ is a symplectic leaf of a Poisson manifold equipped with the structure of a generalized integrable system, see Bottacin [7] and Markman [30].

The pre-image of 0 under the Hitchin map,
$$N = \chi^{-1}(0),$$
is called the *nilpotent cone*. The main result of Laumon [28], proved for the moduli stack of Higgs bundles, is that the nilpotent cone is Lagrangian.

In the non-parabolic case, Hausel [19, Theorem 5.2] proved that the downwards Morse flow on the moduli space of Higgs bundles coincides with the nilpotent cone. His proof goes over word by word to the parabolic case, so we have the following Theorem.

THEOREM 3.13. *The downwards Morse flow on the moduli space of parabolic Higgs bundles coincides with the nilpotent cone N.* □

As pointed out by Hausel, the nilpotent cone is isotropic because the Hitchin map is a completely integrable system. (To be precise, the nilpotent cone being isotropic means that its tangent space at any non-singular point of the nilpotent cone is an isotropic subspace of the tangent space to $\mathcal{M}$, cf. Ginzburg [14].) Hence the nilpotent cone is Lagrangian if its dimension equals half that of the moduli space $\mathcal{M}$. But this fact follows at once from our Corollary 3.10. Thus we have the following version of Laumon's theorem for the moduli space of parabolic Higgs bundles.

THEOREM 3.14. *The nilpotent cone N is a Lagrangian subvariety of the moduli space of parabolic Higgs bundles.* □

CHAPTER 4

Parabolic triples

1. Definitions and basic facts

A *parabolic triple* $T = (E_1, E_2, \phi)$ on X consists of two parabolic vector bundles E_1 and E_2 on X and a $\phi \in H^0(\mathrm{SParHom}(E_2, E_1(D)))$. A homomorphism from $T' = (E'_1, E'_2, \phi')$ to $T = (E_1, E_2, \phi)$ is a commutative diagram

$$\begin{array}{ccc} E'_2 & \xrightarrow{\phi'} & E'_1(D) \\ \downarrow & & \downarrow \\ E_2 & \xrightarrow{\phi} & E_1(D), \end{array}$$

where the vertical arrows are parabolic sheaf homomorphisms. A triple $T' = (E'_1, E'_2, \phi')$ is a subtriple of $T = (E_1, E_2, \phi)$ if the sheaf homomorphims $E'_1 \to E_1$ and $E'_2 \to E_2$ are injective. A subtriple $T' \subset T$ is called *proper* if $T' \neq 0$ and $T' \neq T$.

DEFINITION 4.1. For any $\sigma \in \mathbb{R}$ the σ-degree and σ-slope of T are defined to be

$$\deg_\sigma(T) = \mathrm{pardeg}(E_1) + \mathrm{pardeg}(E_2) + \sigma \ \mathrm{rk}(E_2),$$

$$\mu_\sigma(T) = \frac{\deg_\sigma(T)}{\mathrm{rk}(E_1) + \mathrm{rk}(E_2)}$$

$$= \mathrm{par}\mu(E_1 \oplus E_2) + \sigma \frac{\mathrm{rk}(E_2)}{\mathrm{rk}(E_1) + \mathrm{rk}(E_2)}.$$

We say $T = (E_1, E_2, \phi)$ is σ-stable if

$$\mu_\sigma(T') < \mu_\sigma(T),$$

for any proper subtriple $T' = (E'_1, E'_2, \phi')$. We define σ-semistability by replacing the above strict inequality with a weak inequality. A triple is called σ-*polystable* if it is the direct sum of σ-stable triples of the same σ-slope.

Let us fix the topological and parabolic types of E_1 and E_2. We denote by $\mathcal{N}_\sigma$ the moduli space of σ-stable triples $T = (E_1, E_2, \phi)$ of the given type.

Given a triple $T = (E_1, E_2, \phi)$ one has the dual triple $T^* = (E_2^*, E_1^*, \phi^*)$, where E_i^* is the parabolic dual of E_i and ϕ^* is the transpose of ϕ. The following is not difficult to prove.

PROPOSITION 4.2. *The σ-(semi)stability of the triple T is equivalent to the σ-(semi)stability of the dual triple T^*. The map $T \mapsto T^*$ defines an isomorphism of moduli spaces.* $\square$

This can be used to restrict our study to $\mathrm{rk}(E_1) \geq \mathrm{rk}(E_2)$ and appeal to duality to deal with the case $\mathrm{rk}(E_1) < \mathrm{rk}(E_2)$.

There are certain necessary conditions in order for σ-semistable triples to exist. Let $r_1 = \mathrm{rk}(E_1)$, $r_2 = \mathrm{rk}(E_2)$, $\mathrm{par}\mu_1 = \mathrm{par}\mu(E_1)$ and $\mathrm{par}\mu_2 = \mathrm{par}\mu(E_2)$ be the ranks and parabolic degrees of E_1 and E_2, and define

$$\sigma_m = \mathrm{par}\mu_1 - \mathrm{par}\mu_2, \tag{4.1}$$

$$\sigma_M = \left(1 + \frac{r_1 + r_2}{|r_1 - r_2|}\right)(\mathrm{par}\mu_1 - \mathrm{par}\mu_2) + n\frac{r_1 + r_2}{|r_1 - r_2|}, \quad \text{if } r_1 \neq r_2. \tag{4.2}$$

PROPOSITION 4.3. *A necessary condition for $\mathcal{N}_\sigma$ to be non-empty is*
  (i) $\sigma_m \leq \sigma \leq \sigma_M$, *if* $r_1 \neq r_2$,
  (ii) $\sigma_m \leq \sigma$, *if* $r_1 = r_2$.

PROOF. The proof is similar to the one given in [**8**, Proposition 3.18] for ordinary triples. □

REMARK 4.4. The upper bound given for σ is not optimal. A better one can be found, as will be seen later in Chapter 6.

Using the dimensional reduction construction given in [**4**], the moduli space $\mathcal{N}_\sigma$ can be realised as a subvariety of a certain moduli space of parabolic bundles on $X \times \mathbb{P}^1$. Such moduli spaces have been constructed by Maruyama and Yokogawa [**31**] in arbitrary dimensions using GIT methods.

Another important aspect that follows also from the dimensional reduction point of view is the existence of a correspondence between stability and the existence of solutions to certain gauge-theoretic equations on a parabolic triple $T = (E_1, E_2, \phi)$, known as the *parabolic vortex equations* [**4**]. The parabolic vortex equations

$$\begin{aligned} \mathrm{i}\Lambda F(E_1) + \phi\phi^* &= \tau_1 \,\mathrm{Id}_{E_1}, \\ \mathrm{i}\Lambda F(E_2) - \phi^*\phi &= \tau_2 \,\mathrm{Id}_{E_2}, \end{aligned} \tag{4.3}$$

are equations for Hermitian metrics on E_1 and E_2 adapted to the parabolic structure. Here Λ is contraction by the Kähler form of a metric on X (normalized so that $\mathrm{vol}(X) = 2\pi$), $F(E_i)$ is the curvature of the unique connection on E_i compatible with the Hermitian metric and the holomorphic structure of E_i, and τ_1 and τ_2 are real parameters satisfying $\mathrm{pardeg}(E_1) + \mathrm{pardeg}(E_2) = r_1\tau_1 + r_2\tau_2$. Also, here ϕ^* is the adjoint of ϕ with respect to the Hermitian metrics. One has the following.

THEOREM 4.5. [**4**, Theorem 3.4] *A solution to* (4.3) *exists if and only if T is σ-polystable for $\sigma = \tau_1 - \tau_2$.* □

2. Parabolic Higgs bundles and parabolic triples

The relation between parabolic Higgs bundles and parabolic triples is given by the following.

PROPOSITION 4.6. *Suppose that (E, Φ) is a stable parabolic Higgs bundle such that $E = E_1 \oplus E_2$ and*

$$\Phi = \begin{pmatrix} 0 & \phi \\ 0 & 0 \end{pmatrix}$$

with $\phi : E_2 \to E_1 \otimes K(D)$ a strongly parabolic map. Then (E, Φ) is stable if and only if the parabolic triple $(E_1 \otimes K, E_2, \phi)$ is σ-stable for $\sigma = 2g - 2$.

PROOF. Take a sub-object $E' \subset E$ with $\Phi(E') \subset E' \otimes K(D)$. This can be assumed to be of the form $E' = E'_1 \oplus E'_2$ and hence it defines a subtriple $(E'_1 \otimes K, E'_2, \phi')$ where $\phi' = \phi|_{E'_2}$. The result follows now from the equivalence between

$$\frac{\operatorname{pardeg}(E'_1) + \operatorname{pardeg}(E'_2)}{r'_1 + r'_2} < \frac{\operatorname{pardeg}(E_1) + \operatorname{pardeg}(E_2)}{r_1 + r_2}, \text{ and}$$

$$\frac{\operatorname{pardeg}(E'_1) + r'_1(2g-2) + \operatorname{pardeg}(E'_2)}{r'_1 + r'_2} + \sigma \frac{r'_2}{r'_1 + r'_2} <$$

$$\frac{\operatorname{pardeg}(E_1) + r_1(2g-2) + \operatorname{pardeg}(E_2)}{r_1 + r_2} + \sigma \frac{r_2}{r_1 + r_2},$$

which is the σ-stability of the triple $(E_1 \otimes K, E_2, \Phi)$, for $\sigma = 2g - 2$. □

3. Extensions and deformations of parabolic triples

In order to analyse the differences between the moduli spaces $\mathcal{N}_\sigma$ as σ changes, as well as the smoothness properties of the moduli space for a given value of σ, we need to study the homological algebra of parabolic triples. This is done by considering the hypercohomology of a certain complex of sheaves, in an analogous way to the case of holomorphic triples studied in [**9**], and the parabolic Higgs bundle case studied in Section 2 of Chapter 2.

Let $T' = (E'_1, E'_2, \phi')$ and $T'' = (E''_1, E''_2, \phi'')$ be two parabolic triples. Let $\operatorname{Hom}(T'', T')$ denote the linear space of homomorphisms from T'' to T', and let $\operatorname{Ext}^1(T'', T')$ denote the linear space of equivalence classes of extensions of the form

$$0 \longrightarrow T' \longrightarrow T \longrightarrow T'' \longrightarrow 0,$$

where by this we mean a commutative diagram

$$\begin{array}{ccccccccc}
0 & \longrightarrow & E'_2 & \longrightarrow & E_2 & \longrightarrow & E''_2 & \longrightarrow & 0 \\
& & \phi' \downarrow & & \phi \downarrow & & \phi'' \downarrow & & \\
0 & \longrightarrow & E'_1(D) & \longrightarrow & E_1(D) & \longrightarrow & E''_1(D) & \longrightarrow & 0.
\end{array}$$

Hence, to analyse $\operatorname{Ext}^1(T'', T')$ one considers the complex of sheaves
(4.4)
$$C^\bullet(T'', T') : \operatorname{ParHom}(E''_1, E'_1) \oplus \operatorname{ParHom}(E''_2, E'_2) \xrightarrow{c} \operatorname{SParHom}(E''_2, E'_1(D)),$$

where the map c is defined by

$$c(\psi_1, \psi_2) = \phi'\psi_2 - \psi_1\phi''.$$

PROPOSITION 4.7. *There are natural isomorphisms*

$$\operatorname{Hom}(T'', T') \cong \mathbb{H}^0(C^\bullet(T'', T')),$$

$$\operatorname{Ext}^1(T'', T') \cong \mathbb{H}^1(C^\bullet(T'', T')),$$

and a long exact sequence associated to the complex $C^\bullet(T'', T')$:
(4.5)
$0 \to$
$\mathbb{H}^0 \to H^0(\operatorname{ParHom}(E''_1, E'_1) \oplus \operatorname{ParHom}(E''_2, E'_2)) \to H^0(\operatorname{SParHom}(E''_2, E'_1(D))) \to$
$\mathbb{H}^1 \to H^1(\operatorname{ParHom}(E''_1, E'_1) \oplus \operatorname{ParHom}(E''_2, E'_2)) \to H^1(\operatorname{SParHom}(E''_2, E'_1(D))) \to$
$\mathbb{H}^2 \to 0,$

where $\mathbb{H}^i = \mathbb{H}^i(C^\bullet(T'', T'))$. □

We introduce the following notation:
$$h^i(T'', T') = \dim \mathbb{H}^i(C^\bullet(T'', T')),$$
$$\chi(T'', T') = h^0(T'', T') - h^1(T'', T') + h^2(T'', T').$$

PROPOSITION 4.8. *For any parabolic triples T' and T'' we have*
$$\chi(T'', T') = \chi(\mathrm{ParHom}(E_1'', E_1')) + \chi(\mathrm{ParHom}(E_2'', E_2')) - \chi(\mathrm{SParHom}(E_2'', E_1'(D)))$$
where $\chi(E) = \dim H^0(E) - \dim H^1(E)$ is the Euler characteristic of E.

PROOF. Immediate from the long exact sequence (4.5) and the Riemann–Roch formula. □

COROLLARY 4.9. *For any extension $0 \to T' \to T \to T'' \to 0$ of parabolic triples,*
$$\chi(T, T) = \chi(T', T') + \chi(T'', T'') + \chi(T'', T') + \chi(T', T'').$$
□

PROPOSITION 4.10. *Suppose that T' and T'' are σ-semistable.*
  (i) *If $\mu_\sigma(T') < \mu_\sigma(T'')$ then $\mathbb{H}^0(C^\bullet(T'', T')) \cong 0$.*
  (ii) *If $\mu_\sigma(T') = \mu_\sigma(T'')$ and T' and T'' are both σ-stable, then*
$$\mathbb{H}^0(C^\bullet(T'', T')) \cong \begin{cases} \mathbb{C}, & \text{if } T' \cong T'', \\ 0, & \text{if } T' \not\cong T''. \end{cases}$$
□

COROLLARY 4.11. *Let T' and T'' be σ-semistable parabolic triples with $\mu_\sigma(T') = \mu_\sigma(T'')$, and suppose that $\mathbb{H}^2(C^\bullet(T'', T')) = 0$. Then*
$$\dim \mathrm{Ext}^1(T'', T') = h^0(T'', T') - \chi(T'', T').$$
□

Since the space of infinitesimal deformations of T is isomorphic to $\mathbb{H}^1(C^\bullet(T, T))$, the considerations of the previous sections also apply to studying deformations of a parabolic triple T (the proofs are analogous to the non-parabolic case [**9**]). To be precise, one has the following.

THEOREM 4.12. *Let $T = (E_1, E_2, \phi)$ be a σ-stable parabolic triple.*
  (i) *The Zariski tangent space at the point defined by T in the moduli space of stable triples is isomorphic to $\mathbb{H}^1(C^\bullet(T, T))$.*
  (ii) *If $\mathbb{H}^2(C^\bullet(T, T)) = 0$, then the moduli space of σ-stable parabolic triples is smooth in a neighbourhood of the point defined by T.*
  (iii) *$\mathbb{H}^2(C^\bullet(T, T)) = 0$ if and only if the homomorphism*
$$H^1(\mathrm{ParEnd}(E_1)) \oplus H^1(\mathrm{ParEnd}(E_2)) \longrightarrow H^1(\mathrm{SParHom}(E_2, E_1(D)))$$
   in the corresponding long exact sequence is surjective.
  (iv) *At a smooth point $T \in \mathcal{N}_\sigma$ the dimension of the moduli space of σ-stable parabolic triples is*

(4.6)
$$\dim \mathcal{N}_\sigma = h^1(T, T) = 1 - \chi(T, T)$$
$$= \chi(\mathrm{ParEnd}(E_1, E_1)) + \chi(\mathrm{ParEnd}(E_2, E_2)) - \chi(\mathrm{SParHom}(E_2, E_1(D)))$$

(v) *If ϕ is injective or surjective then $T = (E_1, E_2, \phi)$ defines a smooth point in the moduli space.*

□

CHAPTER 5

Critical values and flips

1. Critical values

A parabolic triple $T = (E_1, E_2, \phi)$ of fixed topological and parabolic type is strictly σ-semistable if and only if it has a proper subtriple $T' = (E_1', E_2', \phi')$ such that $\mu_\sigma(T') = \mu_\sigma(T)$, i.e.

$$\text{par}\mu(E_1' \oplus E_2') + \sigma \frac{r_2'}{r_1' + r_2'} = \text{par}\mu(E_1 \oplus E_2) + \sigma \frac{r_2}{r_1 + r_2}, \tag{5.1}$$

where r_1, r_2, r_1', r_2' are the ranks of E_1, E_2, E_1', E_2'. There are two ways in which this can happen. The first one is if there exists a subtriple T' such that

$$\frac{r_2'}{r_1' + r_2'} = \frac{r_2}{r_1 + r_2}, \text{ and}$$
$$\text{par}\mu(E_1' \oplus E_2') = \text{par}\mu(E_1 \oplus E_2).$$

In this case the terms containing σ drop from (5.1) and T is strictly σ-semistable for all values of σ. We refer to this phenomenon as σ-*independent semistability*.

The other way in which strict σ-semistability can happen is if equality holds in (5.1) but

$$\frac{r_2'}{r_1' + r_2'} \neq \frac{r_2}{r_1 + r_2}.$$

The values of σ for which this happens are called critical values.

From now on we shall make the following assumption on the weights.

ASSUMPTION 5.1. Let $\alpha_i(p)$ be the collection of all the weights of E_1 and E_2 together. We assume that they are all of multiplicity one and that, for a large integer N depending only on the ranks, they satisfy the following property:

$$\sum_{1 \leq i \leq r,\, p \in D} n_{i,p}\, \alpha_i(p) \in \mathbb{Z}, \quad n_{i,p} \in \mathbb{Z}, |n_{i,p}| \leq N \implies n_{i,p} = 0, \text{ for all } p, i.$$

The weights failing this genericity condition are a finite union of hyperplanes in $[0,1)^{nr}$, where nr is the total number of weights, $r = r_1 + r_2$.

PROPOSITION 5.2. (i) *Under Assumption 5.1, there are no σ-independent semistable triples (by taking N larger than $r_1 + r_2$).*
 (ii) *The critical values of σ form a discrete subset of $[\sigma_m, \infty)$, where σ_m is as in (4.1).*
 (iii) *If $r_1 \neq r_2$ the number of critical values is finite and they lie in the interval $[\sigma_m, \sigma_M]$, where σ_M is as in (4.2).*
 (iv) *The stability criteria for two values of σ lying between two consecutive critical values are equivalent; thus the corresponding moduli spaces coincide.*

(v) If σ_c is a critical value and T' is a subtriple of a σ_c-semistable triple T such that $\mu_{\sigma_c}(T') = \mu_{\sigma_c}(T)$, then T' and the quotient triple $T'' = T/T'$ are σ_c-stable (for this, it may be necessary to take a larger value of N in Assumption 5.1).

□

2. Crossing critical values and universal extensions

In this section we study the differences between the moduli spaces $\mathcal{N}_\sigma$, for fixed type but different values of σ.

We begin with a set theoretic description of the differences between two spaces $\mathcal{N}_\sigma$ and $\mathcal{N}_{\sigma'}$ when σ and σ' are separated by a single critical value (as defined in Section 1). For the rest of this chapter we adopt the following notation: when $r_1 \neq r_2$ the bounds σ_m and σ_M are as in (4.1) and (4.2). When $r_1 = r_2$ we adopt the convention that $\sigma_M = \infty$. Let $\sigma_c \in \mathbb{R}$ be a critical value such that
$$\sigma_m \leq \sigma_c \leq \sigma_M.$$
Set
$$\sigma_c^+ = \sigma_c + \epsilon, \quad \sigma_c^- = \sigma_c - \epsilon,$$
where $\epsilon > 0$ is small enough so that σ_c is the only critical value in the interval (σ_c^-, σ_c^+).

DEFINITION 5.3. Let σ_c be a critical value. We define the *flip loci* $\mathcal{S}_{\sigma_c^\pm} \subset \mathcal{N}_{\sigma_c^\pm}$ by the conditions that the points in $\mathcal{S}_{\sigma_c^+}$ represent triples which are σ_c^+-stable but σ_c^--unstable, while the points in $\mathcal{S}_{\sigma_c^-}$ represent triples which are σ_c^--stable but σ_c^+-unstable.

LEMMA 5.4. *In the above notation,*
$$\mathcal{N}_{\sigma_c^+} - \mathcal{S}_{\sigma_c^+} = \mathcal{N}_{\sigma_c} = \mathcal{N}_{\sigma_c^-} - \mathcal{S}_{\sigma_c^-}.$$

□

As a consequence of Proposition 5.2 (v) we have the following.

PROPOSITION 5.5. *Let σ_c be a critical value. Let $T = (E_1, E_2, \phi)$ be a triple of this type which is σ_c-semistable. Then T has a (unique) description as the middle term in an extension*

(5.2) $$0 \to T' \to T \to T'' \to 0$$

in which T' and T'' are σ_c-stable and $\mu_{\sigma_c}(T') = \mu_{\sigma_c}(T) = \mu_{\sigma_c}(T'')$.

□

We thus have the following.

PROPOSITION 5.6. *The set $\mathcal{S}_{\sigma_c^+}$ coincides with the set of equivalence classes of extensions (5.2), in which T' and T'' are σ_c-stable, $\mu_{\sigma_c}(T') = \mu_{\sigma_c}(T) = \mu_{\sigma_c}(T'')$, and $r_2'/r' < r_2''/r''$.*

Similarly, $\mathcal{S}_{\sigma_c^-}$ coincides with the set of equivalence classes of extensions (5.2), in which T' and T'' are σ_c-stable, $\mu_{\sigma_c}(T') = \mu_{\sigma_c}(T) = \mu_{\sigma_c}(T'')$, and $r_2'/r' > r_2''/r''$; or, equivalently, extensions

$$0 \to T'' \to T \to T' \to 0$$

where T' and T'' are as above, but $r_2'/r' < r_2''/r''$.

□

To construct the locus $\mathcal{S}_{\sigma_c^\pm}$, we first observe that, by the genericity of the weights, the moduli spaces $\mathcal{N}'_{\sigma_c}$ and $\mathcal{N}''_{\sigma_c}$ are fine moduli spaces (cf. [**39**]), i.e., there are universal parabolic triples $\mathcal{T}' = (\mathcal{E}'_1, \mathcal{E}'_2, \Phi')$ and $\mathcal{T}'' = (\mathcal{E}''_1, \mathcal{E}''_2, \Phi'')$ over $\mathcal{N}'_{\sigma_c} \times X$ and $\mathcal{N}''_{\sigma_c} \times X$ respectively. Let $B = \mathcal{N}'_{\sigma_c} \times \mathcal{N}''_{\sigma_c}$ and let pull back $\mathcal{T}'$ and $\mathcal{T}''$ to $B \times X$. Considering the complex $C^\bullet(\mathcal{T}'', \mathcal{T}')$ as defined in (4.4), taking relative hypercohomology $\mathbb{H}_\pi(C^\bullet(\mathcal{T}'', \mathcal{T}'))$ with respect to the projection $\pi: B \times X \to B$, and putting

$$(5.3) \qquad W^+ := \mathbb{H}^1_\pi(C^\bullet(\mathcal{T}'', \mathcal{T}')),$$

we have the following exact sequence of sheaves over B:

(5.4)
$$0 \to \mathbb{H}^0_\pi(C^\bullet(\mathcal{T}'', \mathcal{T}')) \to \pi_* \operatorname{ParHom}(\mathcal{E}''_2, \mathcal{E}'_2) \oplus \pi_* \operatorname{ParHom}(\mathcal{E}''_1, \mathcal{E}'_1) \to$$
$$\to \pi_* \operatorname{SParHom}(\mathcal{E}''_2, \mathcal{E}'_1(D)) \to W^+ \to$$
$$\to R^1\pi_* \operatorname{ParHom}(\mathcal{E}''_2, \mathcal{E}'_2) \oplus R^1\pi_* \operatorname{ParHom}(\mathcal{E}''_1, \mathcal{E}'_1) \to R^1\pi_* \operatorname{SParHom}(\mathcal{E}''_2, \mathcal{E}'_1(D))$$
$$\to \mathbb{H}^2_\pi(C^\bullet(\mathcal{T}'', \mathcal{T}')) \to 0.$$

Analogously, we can consider the complex $C^\bullet(\mathcal{T}', \mathcal{T}'')$ and define

$$W^- := \mathbb{H}^1_\pi(C^\bullet(\mathcal{T}', \mathcal{T}'')).$$

PROPOSITION 5.7. *If $\mathbb{H}^2(C^\bullet(\mathcal{T}'', \mathcal{T}')) = 0$ for every $(\mathcal{T}', \mathcal{T}'') \in \mathcal{N}'_{\sigma_c} \times \mathcal{N}''_{\sigma_c}$, then W^+ defined in (5.3) is locally free. Similarly for W^-.*

PROOF. By Proposition 4.10, $\mathbb{H}^0_\pi(C^\bullet(\mathcal{T}'', \mathcal{T}')) = 0$ for every $(\mathcal{T}', \mathcal{T}'') \in \mathcal{N}'_{\sigma_c} \times \mathcal{N}''_{\sigma_c}$ and hence $\mathbb{H}^0_\pi(C^\bullet(\mathcal{T}'', \mathcal{T}')) = 0$. By assumption $\mathbb{H}^2_\pi(C^\bullet(\mathcal{T}'', \mathcal{T}')) = 0$ and the result thus follows from (5.4). $\square$

REMARK 5.8. In our applications, the vanishing assumption in Proposition 5.7 will always be satisfied due to the small rank of the bundles involved. In fact, the vanishing is probably true in general for $\sigma \geq 2g - 2$, as in the non-parabolic case [**9**].

Clearly, from Proposition 5.6, we have the following.

PROPOSITION 5.9. *If $\mathbb{H}^2(C^\bullet(\mathcal{T}'', \mathcal{T}')) = 0$ and $\mathbb{H}^2(C^\bullet(\mathcal{T}', \mathcal{T}'')) = 0$ for every $(\mathcal{T}', \mathcal{T}'') \in \mathcal{N}'_{\sigma_c} \times \mathcal{N}''_{\sigma_c}$, then*

$$\mathcal{S}_{\sigma_c^\pm} = \mathbb{P}W^\pm.$$

$\square$

The following will be important to study the relation between $\mathcal{N}_{\sigma_c^-}$ and $\mathcal{N}_{\sigma_c^+}$.

PROPOSITION 5.10. *Over $\mathbb{P}W^+ \times X$ there is a universal extension*

$$(5.5) \qquad 0 \to \mathcal{T}' \otimes \mathcal{O}_{\mathbb{P}W^+}(1) \to \mathcal{T}^+ \to \mathcal{T}'' \to 0,$$

where $\mathcal{T}' \otimes \mathcal{O}_{\mathbb{P}W^+}(1) := (\mathcal{E}'_1 \otimes \mathcal{O}_{\mathbb{P}W^+}(1), \mathcal{E}'_2 \otimes \mathcal{O}_{\mathbb{P}W^+}(1), \Phi')$ (we omit pull-backs for clarity). Similarly, on $\mathbb{P}W^- \times X$ there is a universal extension

$$0 \to \mathcal{T}'' \otimes \mathcal{O}_{\mathbb{P}W^-}(1) \to \mathcal{T}^- \to \mathcal{T}' \to 0,$$

where $\mathcal{T}'' \otimes \mathcal{O}_{\mathbb{P}W^-}(1) := (\mathcal{E}''_1 \otimes \mathcal{O}_{\mathbb{P}W^-}(1), \mathcal{E}''_2 \otimes \mathcal{O}_{\mathbb{P}W^-}(1), \Phi'')$.

PROOF. The proof is analogous to the one given by Lange [**27**] for extensions of sheaves. In fact, our result could be derived from that one by making use of the correspondence between parabolic triples over X and $SL(2, \mathbb{C})$-invariant parabolic vector bundles over $X \times \mathbb{P}^1$ (cf. [**4**]). Hence we only give the main ingredients of the proof.

Let $(T', T'') \in B = \mathcal{N}'_{\sigma_c} \times \mathcal{N}''_{\sigma_c}$. Let $\mathbb{W} = \mathbb{H}^1(X, C^\bullet(T'', T'))$, and let $\mathbb{P} = \mathbb{P}(\mathbb{W})$. Over $\mathbb{P} \times X$ there is a universal extension

(5.6) $$0 \to T'(1) \to T \to T'' \to 0,$$

where $T'(1) := (E'_1 \otimes \mathcal{O}_{\mathbb{P}}(1), E'_2 \otimes \mathcal{O}_{\mathbb{P}}(1), \phi')$ and we are omitting pull-backs. By the universal property of this extension we mean that T restricted to $\{p\} \times X$ is a triple whose corresponding equivalence class is precisely $p \in \mathbb{P}$. Extensions like (5.6) are parametrised by $\mathbb{H}^1(\mathbb{P} \times X, C^\bullet(T'', T'(1)))$ which by the Künneth formula is isomorphic to

$$\mathbb{H}^1(X, C^\bullet(T'', T')) \otimes H^0(\mathbb{P}, \mathcal{O}_{\mathbb{P}}(1)) \cong \mathbb{W} \otimes \mathbb{W}^* \cong \mathrm{End}(\mathbb{W}).$$

One can show that the identity element in $\mathrm{End}(\mathbb{W})$ defines the universal extension.

To prove the relative version stated in the proposition, we consider the spectral sequence

$$H^p(B, \mathbb{H}^q_\pi(C^\bullet(T'', T'))) \Rightarrow \mathbb{H}^{p+q}(B \times X, C^\bullet(T'', T'))$$

relating relative and global hypercohomology groups. Since $\mathbb{H}^0_\pi(C^\bullet(T'', T')) = 0$, the induced map

$$\mathbb{H}^1(B \times X, C^\bullet(T'', T')) \to H^0(B, \mathbb{H}^1_\pi(C^\bullet(T'', T')))$$

is an isomorphism. Similarly, if $P := \mathbb{P}W^+$, we have an isomorphism

$$\mathbb{H}^1(P \times X, C^\bullet(T'', T' \otimes \mathcal{O}_P(1))) \cong H^0(P, \mathbb{H}^1_\nu(C^\bullet(T'', T' \otimes \mathcal{O}_P(1)))),$$

where $\nu : P \times X \to P$ is the canonical projection.

Now, write $p : P \to B$ for the projection. Again omitting pull-backs when convenient, we have that the image of the identity of W^+ under the canonical isomorphisms

$$\begin{aligned} H^0(B, \mathrm{End}\, W^+) = H^0(B, W^+ \otimes (W^+)^*) &= H^0(B, W^+ \otimes p_* \mathcal{O}_P(1)) \\ &= H^0(B, p_*(p^* W^+ \otimes \mathcal{O}_P(1))) \\ &= H^0(P, p^*(\mathbb{H}^1_\pi(C^\bullet(T'', T'))) \otimes \mathcal{O}_P(1)) \\ &= H^0(P, \mathbb{H}^1_\nu(C^\bullet(T'', T' \otimes \mathcal{O}_P(1)))) \end{aligned}$$

is a nonvanishing section defining the universal extension (5.5). A technical ingredient in proving the universal property is the commutation of $\mathbb{H}^1(B \times X, C^\bullet(T'', T'))$ with base change (see [**27**] for details on the analogous situation of extensions of sheaves). □

3. Flips

Now we assume that $\mathcal{N}_{\sigma_c^+}$, $\mathcal{N}_{\sigma_c^-}$, $\mathcal{N}'_{\sigma_c}$ and $\mathcal{N}''_{\sigma_c}$ are smooth. Also assume that $\mathbb{H}^2(C^\bullet(T'', T')) = 0$ and $\mathbb{H}^2(C^\bullet(T', T'')) = 0$ for every $(T', T'') \in \mathcal{N}'_{\sigma_c} \times \mathcal{N}''_{\sigma_c}$. This will always be the case in our applications. In order to relate $\mathcal{N}_{\sigma_c^-}$ and $\mathcal{N}_{\sigma_c^+}$ we have to blow up $\mathcal{N}_{\sigma_c^\pm}$ along $\mathcal{S}_{\sigma_c^\pm}$. For this it is necessary to study the normal bundle to $\mathcal{S}_{\sigma_c^\pm} = \mathbb{P}W^\pm$ in $\mathcal{N}_{\sigma_c^\pm}$.

PROPOSITION 5.11. *Let $p : \mathbb{P}W^\pm \to B$ be the natural projection and $j : \mathbb{P}W^\pm \hookrightarrow \mathcal{N}_{\sigma_c^\pm}$ be the natural inclusion. Then there is an exact sequence*

$$0 \longrightarrow T\mathbb{P}W^\pm \longrightarrow j^*T\mathcal{N}_{\sigma_c^\pm} \longrightarrow p^*W^\mp \otimes \mathcal{O}_{\mathbb{P}W^\pm}(-1) \longrightarrow 0,$$

and hence, the normal bundle to $\mathcal{S}_{\sigma_c^\pm} = \mathbb{P}W^\pm$ in $\mathcal{N}_{\sigma_c^\pm}$ is isomorphic to $p^*W^\mp \otimes \mathcal{O}_{\mathbb{P}W^\pm}(-1)$.

PROOF. We consider the case of $\mathcal{N}_{\sigma_c^+}$ — the case of $\mathcal{N}_{\sigma_c^-}$ is analogous. Over $\mathcal{N}_{\sigma_c^+} \times X$ there is a universal triple $\mathcal{T} = (\mathcal{E}_1, \mathcal{E}_2, \Phi)$, whose restriction to $\mathcal{S}_{\sigma_c^+} = \mathbb{P}W^+$ is the universal extension $\mathcal{T}^+$ in (5.5). The tangent bundle of $\mathcal{N}_{\sigma_c^+}$ is given by the relative $\mathbb{H}^1$ of the complex

$$\mathrm{ParHom}(\mathcal{E}_1, \mathcal{E}_1) \oplus \mathrm{ParHom}(\mathcal{E}_2, \mathcal{E}_2) \longrightarrow \mathrm{SParHom}(\mathcal{E}_2, \mathcal{E}_1(D))$$

over $\mathcal{N}_{\sigma_c^+} \times X$ with respect to the natural projection $\mathcal{N}_{\sigma_c^+} \times X \to \mathcal{N}_{\sigma_c^+}$.

Denote $\mathcal{E}'_i(1) = \mathcal{E}'_i \otimes \mathcal{O}_{\mathbb{P}W^+}(1)$ and define

$$\mathrm{ParHom}_U(\mathcal{E}_i, \mathcal{E}_i) := \ker \left(\mathrm{ParHom}(\mathcal{E}_i, \mathcal{E}_i) \to \mathrm{ParHom}(\mathcal{E}'_i(1), \mathcal{E}''_i) \right)$$

and

$$\mathrm{SParHom}_U(\mathcal{E}_2, \mathcal{E}_1(D)) := \ker \left(\mathrm{SParHom}(\mathcal{E}_2, \mathcal{E}_1(D)) \to \mathrm{SParHom}(\mathcal{E}'_2(1), \mathcal{E}''_1(D)) \right).$$

The tangent bundle of $\mathbb{P}W^+$ is the relative $\mathbb{H}^1$ with respect to the projection $\mathbb{P}W^+ \times X \to \mathbb{P}W^+$ of the middle complex in the following exact sequence of complexes

$$\begin{array}{ccc}
\mathrm{ParHom}(\mathcal{E}''_1, \mathcal{E}'_1(1)) \oplus \mathrm{ParHom}(\mathcal{E}''_2, \mathcal{E}'_2(1)) & \longrightarrow & \mathrm{SParHom}(\mathcal{E}''_2, \mathcal{E}'_1(1)(D)) \\
\downarrow & & \downarrow \\
\mathrm{ParHom}_U(\mathcal{E}_1, \mathcal{E}_1) \oplus \mathrm{ParHom}_U(\mathcal{E}_2, \mathcal{E}_2) & \longrightarrow & \mathrm{SParHom}_U(\mathcal{E}_2, \mathcal{E}_1(D)) \\
\downarrow & & \downarrow \\
\begin{array}{c}\mathrm{ParHom}(\mathcal{E}'_1(1), \mathcal{E}'_1(1)) \oplus \mathrm{ParHom}(\mathcal{E}'_2(1), \mathcal{E}'_2(1)) \\ \oplus \mathrm{ParHom}(\mathcal{E}''_1, \mathcal{E}''_1) \oplus \mathrm{ParHom}(\mathcal{E}''_2, \mathcal{E}''_2)\end{array} & \longrightarrow & \begin{array}{c}\mathrm{SParHom}(\mathcal{E}'_2(1), \mathcal{E}'_1(1)(D)) \\ \oplus \mathrm{SParHom}(\mathcal{E}''_2, \mathcal{E}''_1(D)).\end{array}
\end{array}$$

Note that when passing to cohomology, this gives us the exact sequence

$$0 \longrightarrow T_V \mathbb{P}W^+ \longrightarrow T\mathbb{P}W^+ \longrightarrow T(\mathcal{N}'_{\sigma_c} \times \mathcal{N}''_{\sigma_c}) \longrightarrow 0,$$

where $T_V \mathbb{P}W^+ \cong p^*W^+(1)/\mathcal{O}_{\mathbb{P}W^+}$ is the vertical tangent bundle.

Therefore the normal bundle to $\mathbb{P}W^+$ is the relative $\mathbb{H}^1$ of the quotient complex in the following exact sequence of complexes

$$\begin{array}{ccc}
\mathrm{ParHom}_U(\mathcal{E}_1, \mathcal{E}_1) \oplus \mathrm{ParHom}_U(\mathcal{E}_2, \mathcal{E}_2) & \longrightarrow & \mathrm{SParHom}_U(\mathcal{E}_2, \mathcal{E}_1(D)) \\
\downarrow & & \downarrow \\
\mathrm{ParHom}(\mathcal{E}_1, \mathcal{E}_1) \oplus \mathrm{ParHom}(\mathcal{E}_2, \mathcal{E}_2) & \longrightarrow & \mathrm{SParHom}(\mathcal{E}_2, \mathcal{E}_1(D)) \\
\downarrow & & \downarrow \\
\mathrm{ParHom}(\mathcal{E}'_1(1), \mathcal{E}''_1) \oplus \mathrm{ParHom}(\mathcal{E}'_2(1), \mathcal{E}''_2) & \longrightarrow & \mathrm{SParHom}(\mathcal{E}'_2(1), \mathcal{E}''_1(D)),
\end{array}$$

and it is hence isomorphic to $p^*W^- \otimes \mathcal{O}_{\mathbb{P}W^+}(-1)$. □

In particular, we conclude that the embedding $\mathbb{P}W^\pm \hookrightarrow \mathcal{N}_{\sigma_c^\pm}$ is smooth. So we can blow-up $\mathcal{N}_{\sigma_c^\pm}$ along $\mathbb{P}W^\pm$ to get $\widetilde{\mathcal{N}}_{\sigma_c^\pm}$ with exceptional divisor $E_\pm \subset \widetilde{\mathcal{N}}_{\sigma_c^\pm}$ such that

$$E_\pm = \mathbb{P}W^\pm \times_B \mathbb{P}W^\mp.$$

Note that $\mathcal{O}_{E_\pm}(E_\pm) = \mathcal{O}_{\mathbb{P}W^+}(-1) \otimes \mathcal{O}_{\mathbb{P}W^-}(-1)$, by Proposition 5.11.

PROPOSITION 5.12. *There is a natural isomorphism $\widetilde{\mathcal{N}}_{\sigma_c^+} \cong \widetilde{\mathcal{N}}_{\sigma_c^-}$.*

PROOF. Let $\mathcal{T}$ be the universal triple over $\mathcal{N}_{\sigma_c^+} \times X$. By Proposition 5.10, the restriction of $\mathcal{T}$ to $\mathbb{P}W^+ \times X$ lies in the universal extension

$$(5.7) \qquad 0 \to \mathcal{T}' \otimes \mathcal{O}_{\mathbb{P}W^+}(1) \to \mathcal{T}|_{\mathbb{P}W^+ \times X} \to \mathcal{T}'' \to 0.$$

Now pull back $\mathcal{T}$ by the blow-up map $q : \widetilde{\mathcal{N}}_{\sigma_c^+} \to \mathcal{N}_{\sigma_c^+}$. Consider the composition $q^*\mathcal{T} \to q^*\mathcal{T}|_{E_+ \times X} \to q^*i_*\mathcal{T}''$, where $i : \mathbb{P}W^+ \times X \hookrightarrow \mathcal{N}_{\sigma_c^+} \times X$. Since $q^*i_*\mathcal{T}''$ is a triple formed by two bundles supported on a divisor, the kernel is a triple (i.e. it is formed by two bundles, and not just coherent sheaves). Define the triple $\hat{\mathcal{T}}$ on $\widetilde{\mathcal{N}}_{\sigma_c^+} \times X$ by the exact sequence

$$(5.8) \qquad 0 \longrightarrow \hat{\mathcal{T}} \otimes \mathcal{O}_{\mathbb{P}W^+}(1) \longrightarrow q^*\mathcal{T} \longrightarrow q^*i_*\mathcal{T}'' \longrightarrow 0.$$

(This is called an *elementary transformation*.)

Let us see that all the triples in the family $\hat{\mathcal{T}}$ are σ_c^--stable. Therefore this defines a map $\widetilde{\mathcal{N}}_{\sigma_c^+} \to \mathcal{N}_{\sigma_c^-}$. Obviously, off $E_+ \times X$, $\hat{\mathcal{T}} \cong \mathcal{T}$ is the family parametrising triples which are σ_c^+-stable and σ_c^--stable at the same time. Tensoring the exact sequence (5.8) with $\mathcal{O}_{E_+ \times X}$, we obtain, over $E_+ \times X$,

$$0 \to \underline{\mathrm{Tor}}(q^*i_*\mathcal{T}'', \mathcal{O}_{E_+ \times X}) \to \hat{\mathcal{T}}|_{E_+ \times X} \otimes \mathcal{O}_{\mathbb{P}W^+}(1) \to q^*\mathcal{T}|_{E_+ \times X} \to q^*i_*\mathcal{T}'' \to 0.$$

Since $\underline{\mathrm{Tor}}(q^*i_*\mathcal{T}'', \mathcal{O}_{E_+ \times X}) = \mathcal{T}'' \otimes \mathcal{O}_{E_+ \times X}(-E_+ \times X) = \mathcal{T}'' \otimes \mathcal{O}_{\mathbb{P}W^-}(1) \otimes \mathcal{O}_{\mathbb{P}W^+}(1)$, and also using (5.7), we get a triple

$$(5.9) \qquad 0 \to \mathcal{T}'' \otimes \mathcal{O}_{\mathbb{P}W^-}(1) \to \hat{\mathcal{T}}|_{E_+ \times X} \to \mathcal{T}' \to 0.$$

We have to check that all extensions in the family (5.9) are non-trivial. For this, restrict to the fibre $\mathbb{P}W^+ \times \mathbb{P}W^-$ over a point $b \in B$. This corresponds to fixing some specific triples T' and T''. We have an exact sequence

$$0 \to T'' \otimes \mathcal{O}_{\mathbb{P}W_b^-}(1) \to \hat{\mathcal{T}}|_{\mathbb{P}W_b^+ \times \mathbb{P}W_b^- \times X} \to T' \to 0.$$

This extension class is parametrised by

$$\mathrm{Ext}^1(T', T'' \otimes \mathcal{O}_{\mathbb{P}W_b^-}(1)) = W_b^- \otimes H^0(\mathcal{O}_{\mathbb{P}W_b^-}(1)) = \mathrm{End}(W_b^-).$$

Moreover the linear group $GL(W_b^-)$ acts on W_b^-. The extension class is invariant by this action, therefore it is a linear multiple of the identity. Letting b move in B we have a section of $\mathrm{End}(W^-)$. Since this is a multiple of the identity, it lives in $\mathcal{O} \cdot \mathrm{Id} \subset \mathrm{End}(W)$, therefore it is a *constant* multiple of the identity. This cannot be constantly zero for, otherwise, it would be $\hat{\mathcal{T}}|_{E_+ \times X} = \mathcal{T}' \oplus (\mathcal{T}'' \otimes \mathcal{O}_{\mathbb{P}W^-}(1))$. Then $\mathrm{Hom}(\hat{\mathcal{T}}, q^*i_*\mathcal{T}'' \otimes \mathcal{O}_{\mathbb{P}W^-}(1)) \neq 0$. Hence (5.8) would imply that the map
$$(5.10)$$
$$\mathrm{Ext}^1(\mathcal{T}'' \otimes \mathcal{O}_{\mathbb{P}W^+}(-1), \mathcal{T}'' \otimes \mathcal{O}_{\mathbb{P}W^-}(1)) \to \mathrm{Ext}^1(\mathcal{T} \otimes \mathcal{O}_{\mathbb{P}W^+}(-1), \mathcal{T}'' \otimes \mathcal{O}_{\mathbb{P}W^-}(1))$$

is not injective. On the other hand, using the proyection $\pi : \mathcal{N}_{\sigma_c^+} \times X \to \mathcal{N}_{\sigma_c^+}$ in (5.7) we have that

$$\mathrm{Hom}_\pi(\mathcal{T}'', \mathcal{T}'') \cong \mathrm{Hom}_\pi(\mathcal{T}, \mathcal{T}''),$$
$$\mathrm{Ext}^1_\pi(\mathcal{T}'', \mathcal{T}'') \hookrightarrow \mathrm{Ext}^1_\pi(\mathcal{T}, \mathcal{T}''),$$

as bundles over $\mathcal{N}_{\sigma_c^+}$. Twisting by $\mathcal{O}_{\mathbb{P}W^+}(1) \otimes \mathcal{O}_{\mathbb{P}W^-}(1)$ and using the spectral sequence

$$H^p(\mathcal{N}_{\sigma_c^+}, \mathrm{Ext}^q_\pi(\cdot, \cdot)) \Rightarrow \mathbb{H}^{p+q}(\mathcal{N}_{\sigma_c^+} \times X, C^\bullet(\cdot, \cdot))$$

we have that (5.10) is injective, giving a contradiction.

Hence the extension class of (5.9) is a non-zero multiple of the identity. This gives a map $\tilde{\mathcal{N}}_{\sigma_c^+} \to \mathcal{N}_{\sigma_c^-}$ which restricts to E_+ as the natural projection on the second factor $E_+ \cong \mathbb{P}W^+ \times_B \mathbb{P}W^- \to \mathbb{P}W^-$. Analogously we obtain a map $\tilde{\mathcal{N}}_{\sigma_c^-} \to \mathcal{N}_{\sigma_c^+}$. So there are two *injective* maps $\tilde{\mathcal{N}}_{\sigma_c^\pm} \to \mathcal{N}_{\sigma_c^+} \times \mathcal{N}_{\sigma_c^-}$. Their images are both the closures of the image of $\tilde{\mathcal{N}}_{\sigma_c^\pm} \smallsetminus E_\pm$, which are the same. So they coincide. □

REMARK 5.13. Let σ_c be a critical value. If $w^+ = \mathrm{rk}(W^+) > 0$ and $w^- = \mathrm{rk}(W^-) > 0$ then the moduli spaces $\mathcal{N}_{\sigma_c^-}$ and $\mathcal{N}_{\sigma_c^+}$ are birational. This is true because $w^+ + w^- + \dim B - 1 = \dim \mathcal{N}_{\sigma_c^\pm}$ by Corollary 4.9, and the flip loci $\mathcal{S}_{\sigma_c^\pm}$ have dimension $w^\pm + \dim B - 1$.

CHAPTER 6

Parabolic triples with $r_1 = 2$ and $r_2 = 1$

In this chapter we use the results of Chapter 5 to compute the Poincaré polynomial of the moduli space of parabolic triples $\mathcal{N}_\sigma$ for the case $r_1 = 2$ and $r_2 = 1$ and for non-critical values of σ. We are studying triples of the form $\phi : L \to E_1(D)$, where L is a parabolic line bundle with $\deg(L) = d_2$ and weights $\alpha(p)$, and E_1 is a parabolic rank 2 bundle with $\deg(E_1) = d_1$ and weights $\beta_1(p) < \beta_2(p)$. By Theorem 4.12 (v), the moduli space of stable elements in $\mathcal{N}_\sigma$ is smooth. Moreover, applying the exact sequence (4.5), one can easily show that we are in the situation given in Proposition 5.7.

1. Flips

By Section 1 of Chapter 5, there are the following three possibilities for the existence of critical values:

- $r_1' = 1$ and $r_2' = 0$. Then the subtriple T' is of the form $0 \to M(D)$, where M is a line bundle of degree d_M. Since M inherits weights from E_1, there is a function $\varepsilon = \{\varepsilon(p)\}_{p \in D}$, which assigns to each $p \in D$ a number $\varepsilon(p) \in \{1, 2\}$ such that the weight of M at p is $\beta_{\varepsilon(p)}(p)$. We have an exact sequence of triples

$$\begin{array}{ccccc} 0 & \longrightarrow & L & \longrightarrow & L \\ \downarrow & & \downarrow & & \downarrow \\ M(D) & \longrightarrow & E_1(D) & \longrightarrow & F(D) \,. \end{array}$$

The quotient triple is of the form $L \to F(D)$, where F is a parabolic line bundle of degree $d_1 - d_M$ and weights $\beta_{\varsigma(p)}(p)$, with $\varsigma(p) = 3 - \varepsilon(p)$. Note that $\{\beta_{\varepsilon(p)}(p), \beta_{\varsigma(p)}(p)\} = \{\beta_1(p), \beta_2(p)\}$. By (5.1), the critical value is

(6.1) $$\sigma_c = 3d_M - d_1 - d_2 + \sum_p \left(2\beta_{\varepsilon(p)}(p) - \alpha(p) - \beta_{\varsigma(p)}(p)\right).$$

As described in Section 3 of Chapter 5, this defines the subspace $\mathcal{S}_{\sigma_c+} = \mathbb{P}W_{\sigma_c}^+$, where

$$W_{\sigma_c}^+ \longrightarrow B_{\sigma_c} = \mathcal{N}_{\sigma_c}' \times \mathcal{N}_{\sigma_c}''$$

(we shall make the dependence on σ_c explicit in this section, since we shall be working with various flip loci simultaneously). The moduli space parametrizing the possible parabolic line bundles M with fixed weights $\beta_{\varepsilon(p)}(p)$ is $\mathcal{N}_{\sigma_c}' = \mathrm{Jac}^{d_M} X$. The moduli space parametrizing triples of the form $L \to F(D)$, which are parabolic line bundles with fixed weights is $\mathrm{Jac}^{d_2} X \times S^N X$, where $N = \deg \mathrm{SParHom}(L, F(D))$. To compute this we use the following.

33

6. PARABOLIC TRIPLES WITH $r_1 = 2$ AND $r_2 = 1$

LEMMA 6.1. *Let L_1, L_2 be two parabolic line bundles with weights $\alpha_{L_1}(p)$ and $\alpha_{L_2}(p)$, respectively. Then*

$$\mathrm{SParHom}(L_1, L_2 \otimes K(D)) \cong \mathrm{Hom}(L_1, L_2 \otimes K(S)),$$

where $S = \{p \in D \mid \alpha_{L_1}(p) < \alpha_{L_2}(p)\}$.

PROOF. By definition, a strongly parabolic map $\Phi : L_1 \to L_2$ satisfies

$$\mathrm{Res}_p \Phi = 0 \iff \alpha_{L_1}(p) \geq \alpha_{L_2}(p).$$

From this the result is clear. □

In our case, we introduce the following notations:

(6.2)
$$\begin{aligned} S_1 &= \{p \in D \mid \alpha(p) < \beta_{\varsigma(p)}(p)\}, \\ S_2 &= \{p \in D \mid \alpha(p) < \beta_{\varepsilon(p)}(p)\}, \\ S_3 &= \{p \in D \mid \beta_{\varepsilon(p)}(p) < \beta_{\varsigma(p)}(p)\}, \\ s_1 &= \#S_1, \ s_2 = \#S_2, \ s_3 = \#S_3. \end{aligned}$$

Then

(6.3)
$$\begin{aligned} N &= \deg \mathrm{SParHom}(L, F(D)) = \deg \mathrm{Hom}(L, F(S_1)) = \\ &= \deg(F) - \deg(L) + s_1 = d_1 - d_2 - d_M + s_1. \end{aligned}$$

Now $\mathbb{P}W_{\sigma_c}^+$ is a projective fibration over B_{σ_c} with fibres projective spaces of dimension $w_{\sigma_c}^+ - 1$. By Proposition 4.8 and Corollary 4.11,

(6.4)
$$\begin{aligned} w_{\sigma_c}^+ &= \dim \mathrm{Ext}^1(T'', T') = -\chi(T'', T') \\ &= -\chi(\mathrm{ParHom}(F, M)) + \chi(\mathrm{SParHom}(L, M(D))) \\ &= -\chi(\mathrm{Hom}(F, M(-S_3))) + \chi(\mathrm{Hom}(L, M(S_2))) \\ &= d_1 - d_2 - d_M + s_2 + s_3. \end{aligned}$$

- $r_1' = 1$, $r_2' = 1$. Then the subtriple T' is of the form $L \to F(D)$ and the quotient triple is of the form $0 \to M(D)$, yielding an exact sequence

$$\begin{array}{ccccc} L & \longrightarrow & L & \longrightarrow & 0 \\ \downarrow & & \downarrow & & \downarrow \\ F(D) & \longrightarrow & E(D) & \longrightarrow & M(D), \end{array}$$

where M is a line bundle of degree d_M and weights $\beta_{\varepsilon(p)}(p)$, for some $\varepsilon = \{\varepsilon(p)\}_{p \in D}$, and F is a parabolic line bundle of degree $d_1 - d_M$ and weights $\beta_{\varsigma(p)}(p)$, with $\varsigma(p) = 3 - \varepsilon(p)$. The critical value is again given by (6.1). These extensions define the subspace $\mathcal{S}_{\sigma_c^-} = \mathbb{P}W_{\sigma_c}^-$, where

$$W_{\sigma_c}^- \longrightarrow B_{\sigma_c} = \mathcal{N}'_{\sigma_c} \times \mathcal{N}''_{\sigma_c} = \mathrm{Jac}^{d_M} X \times \mathrm{Jac}^{d_2} X \times S^N X,$$

with N as in (6.3). Now $\mathbb{P}W_{\sigma_c}^-$ is a projective fibration over B_{σ_c} with fibres projective spaces of dimension $w_{\sigma_c}^- - 1$ where

(6.5)
$$\begin{aligned} w_{\sigma_c}^- &= h^1(T', T'') = -\chi(T', T'') = -\chi(\mathrm{ParHom}(M, F)) = \\ &= -\chi(\mathrm{Hom}(M, F(-(D - S_3)))) = 2d_M - d_1 + g - 1 + n - s_3. \end{aligned}$$

- $r_1' = 2$, $r_2' = 0$. Then the triples T are extensions of $0 \to E_1$ by $L \to 0$. The critical value is $\sigma_c = \mathrm{par}\mu_1 - \mathrm{par}\mu_2 = \sigma_m$, which is the minimum possible value for the parameter σ. At this value, the moduli space $\mathcal{N}_{\sigma_m^-} = \emptyset$ and $\mathcal{N}_{\sigma_m^+} = \mathcal{S}_{\sigma_m^+}$. This can be described explicitly as a projective fibration over a product of a Jacobian and a moduli space of rank 2 stable parabolic bundles, but we will not go into this since we shall not use it.

REMARK 6.2. Since we are taking generic values for the weights, the values of σ_c are distinct for the different choices of d_M and ε. The genericity condition was necessary in Chapter 5 to have smooth flip loci, and this is essentially due to the fact that at a critical value, the Jordan-Hölder filtrations are of length at most two. In the case we treat here, $r_1 = 2$ and $r_2 = 1$, the Jordan-Hölder filtrations are of length at most two even with non-generic weights, because the ranks are too small. Therefore the computations of this chapter work as well for the case of *distinct* non-generic weights. Of course, we shall need the genericity of weights at many other places in the coming chapters.

REMARK 6.3. Let σ_L be the largest critical value. This means that the moduli space $\mathcal{N}_{\sigma_L^+} = \emptyset$, i.e., $\mathcal{S}_{\sigma_L^-} = \mathcal{N}_{\sigma_L^-}$. Since

$$\dim B_{\sigma_c} + w_{\sigma_c}^+ + w_{\sigma_c}^- - 1 = \dim \mathcal{N}_{\sigma_c^\pm},$$

by Corollary 4.9, we have that $w_{\sigma_L}^+ = 0$. By (6.4), this corresponds to the case $s_3 = 0$ and $d_1 - d_2 - d_M + s_2 = 0$, i.e., when $L \cong F(S_2) \subset F(D)$. In this case $\mathcal{N}_{\sigma_L^-}$ equals $\mathbb{P}W_{\sigma_L}^- \to B_{\sigma_L}$.

Also $\sigma_L < \sigma_M$; in general they are not equal. The value of σ_L obtained in (6.1) is always slightly smaller than that of σ_M in (4.2).

2. Poincaré polynomial of moduli of triples

Let σ_c be a critical value as in Section 1 with the only condition $\sigma_c \neq \sigma_m$. Then we have that

(6.6) $$P_t(\mathcal{N}_{\sigma_c^-}) - P_t(\mathcal{N}_{\sigma_c^+}) = P_t(\mathbb{P}W_{\sigma_c}^-) - P_t(\mathbb{P}W_{\sigma_c}^+).$$

Note that this formula also holds when $w_{\sigma_c}^+ = 0$ or $w_{\sigma_c}^- = 0$. For instance, if $w_{\sigma_c}^+ = 0$ then $\mathcal{S}_{\sigma_c^+} = \emptyset$ and $\mathcal{S}_{\sigma_c^-} = \mathbb{P}W_{\sigma_c}^-$ is of the same dimension as $\mathcal{N}_{\sigma_c^-}$, hence it is a component of it. So (6.6) holds. In particular we can use (6.6) for $\sigma_c = \sigma_L$ (see Remark 6.3). But we cannot use it for $\sigma_c = \sigma_m$ (see Remark 6.5).

THEOREM 6.4. *Let $\sigma > \sigma_m$ be a non-critical value. For any $\varepsilon = \{\varepsilon(p)\}_{p \in D}$, $\varepsilon(p) \in \{1,2\}$, let s_1, s_2, s_3 be given by (6.2) and*

$$\bar{d}_M = \left[\frac{1}{3}\left(d_1 + d_2 + \sum \left(\alpha(p) + \beta_{\varsigma(p)}(p) - 2\beta_{\varepsilon(p)}(p)\right) + \sigma\right)\right] + 1,$$

where $\varsigma(p) = 3 - \varepsilon(p)$ and $[x]$ is the integer part of x. Then $P_t(\mathcal{N}_\sigma)$ equals

$$\sum_\varepsilon \mathop{\mathrm{Coeff}}_{x^0} \left(\frac{(1+t)^{4g}(1+xt)^{2g} t^{2d_1 - 2d_2 + 2s_2 + 2s_3 - 2\bar{d}_M} x^{\bar{d}_M}}{(1-t^2)(1-x)(1-xt^2)(1-t^{-2}x) x^{d_1 - d_2 + s_1}} \right.$$

$$\left. - \frac{(1+t)^{4g}(1+xt)^{2g} t^{-2d_1 + 2g - 2 + 2n - 2s_3 + 4\bar{d}_M} x^{\bar{d}_M}}{(1-t^2)(1-x)(1-xt^2)(1-t^4 x) x^{d_1 - d_2 + s_1}} \right).$$

PROOF. From (6.6), we have that

$$
\begin{aligned}
P_t(\mathcal{N}_\sigma) &= \sum_{\sigma_c > \sigma} \left(P_t(\mathbb{P}W_{\sigma_c}^-) - P_t(\mathbb{P}W_{\sigma_c}^+) \right) \\
&= \sum_{\sigma_c > \sigma} \left(P_t(\mathbb{P}^{w_{\sigma_c}^- - 1}) - P_t(\mathbb{P}^{w_{\sigma_c}^+ - 1}) \right) P_t(B_{\sigma_c}) \\
&= \sum_{\sigma_c > \sigma} \left(\frac{1 - t^{2w_{\sigma_c}^-}}{1 - t^2} - \frac{1 - t^{2w_{\sigma_c}^+}}{1 - t^2} \right) P_t(\operatorname{Jac} X)^2 P_t(\operatorname{Sym}^N X) \\
&= \sum_{\sigma_c > \sigma} \frac{t^{2w_{\sigma_c}^+} - t^{2w_{\sigma_c}^-}}{1 - t^2} (1 + t)^{4g} \underset{x^0}{\operatorname{Coeff}} \left(\frac{(1 + xt)^{2g}}{(1 - x)(1 - xt^2)x^N} \right) \\
&= \sum_{\sigma_c > \sigma} \frac{t^{2d_1 - 2d_2 - 2d_M + 2s_2 + 2s_3} - t^{4d_M - 2d_1 + 2g - 2 + 2n - 2s_3}}{1 - t^2} (1 + t)^{4g} \cdot \\
&\quad \cdot \underset{x^0}{\operatorname{Coeff}} \left(\frac{(1 + xt)^{2g}}{(1 - x)(1 - xt^2)x^{d_1 - d_2 - d_M + s_1}} \right) \\
&= \sum_\varepsilon \underset{x^0}{\operatorname{Coeff}} \left(\frac{(1 + t)^{4g}(1 + xt)^{2g} t^{2d_1 - 2d_2 + 2s_2 + 2s_3}}{(1 - t^2)(1 - x)(1 - xt^2) x^{d_1 - d_2 + s_1}} \sum_{d_M | \sigma_c > \sigma} t^{-2d_M} x^{d_M} \right. \\
&\quad \left. - \frac{(1 + t)^{4g}(1 + xt)^{2g} t^{-2d_1 + 2g - 2 + 2n - 2s_3}}{(1 - t^2)(1 - x)(1 - xt^2) x^{d_1 - d_2 + s_1}} \sum_{d_M | \sigma_c > \sigma} t^{4d_M} x^{d_M} \right).
\end{aligned}
$$

We have used [29] in the fourth line, and the fifth line follows from (6.3), (6.4) and (6.5).

The condition for d_M is

$$\sigma_c = 3d_M - d_1 - d_2 + \sum \left(2\beta_{\varepsilon(p)}(p) - \alpha(p) - \beta_{\varsigma(p)}(p) \right) > \sigma ,$$

which translates into

$$d_M > \frac{1}{3} \left(d_1 + d_2 + \sum \left(\alpha(p) + \beta_{\varsigma(p)}(p) - 2\beta_{\varepsilon(p)}(p) \right) + \sigma \right).$$

Since σ is not a critical value, we cannot have equality, so the right hand side is not an integer. The inequality becomes $d_M \geq \bar{d}_M$, with $\bar{d}_M$ as in the statement. Now

$$
\begin{aligned}
\sum_{d_M = \bar{d}_M}^\infty t^{-2d_M} x^{d_M} &= \frac{t^{-2\bar{d}_M} x^{\bar{d}_M}}{1 - t^{-2} x} , \\
\sum_{d_M = \bar{d}_M}^\infty t^{4d_M} x^{d_M} &= \frac{t^{4\bar{d}_M} x^{\bar{d}_M}}{1 - t^4 x} .
\end{aligned}
$$

So finally

$$
\begin{aligned}
P_t(\mathcal{N}_\sigma) &= \sum_\varepsilon \underset{x^0}{\operatorname{Coeff}} \left(\frac{(1 + t)^{4g}(1 + xt)^{2g} t^{2d_1 - 2d_2 + 2s_2 + 2s_3} t^{-2\bar{d}_M} x^{\bar{d}_M}}{(1 - t^2)(1 - x)(1 - xt^2)(1 - t^{-2} x) x^{d_1 - d_2 + s_1}} \right. \\
&\quad \left. - \frac{(1 + t)^{4g}(1 + xt)^{2g} t^{-2d_1 + 2g - 2 + 2n - 2s_3} t^{4\bar{d}_M} x^{\bar{d}_M}}{(1 - t^2)(1 - x)(1 - xt^2)(1 - t^4 x) x^{d_1 - d_2 + s_1}} \right).
\end{aligned}
$$

□

REMARK 6.5. The formula in this theorem only works for $\sigma > \sigma_m$. For $\sigma < \sigma_m$, $\mathcal{N}_\sigma$ is empty, but the formula above does not give zero for such values.

CHAPTER 7

Critical submanifolds of type $(1,1,1)$

1. Description of the critical submanifolds

In this chapter we consider the critical points of the Bott–Morse function f represented by parabolic Higgs bundles (E, Φ) of type $(1,1,1)$, i.e., of the form $E = L_1 \oplus L_2 \oplus L_3$ where L_l are parabolic line bundles, i.e., line bundles with weights at the points $p \in D$. We denote the (fixed) weights of (E, Φ) at $p \in D$ by $0 \leq \alpha_1(p) < \alpha_2(p) < \alpha_3(p) < 1$. Each possible choice of the distribution of these weights among the line bundles L_l is given by a permutation $\varpi_p \in S_3$ such that the weight on the fibre $L_{l,p}$ at p is $\alpha_{\varpi_p(l)}(p) = \alpha_{\varpi(l)}(p)$ for $l = 1, 2, 3$. The map Φ decomposes as strongly parabolic maps $\Phi_1 : L_1 \to L_2 \otimes K(D)$ and $\Phi_2 : L_2 \to L_3 \otimes K(D)$.

We define

$$d_l = \deg(L_l) \quad \text{for } l = 1, 2, 3,$$
$$m = d_1 + d_2.$$

We shall choose to describe the topological data (d_1, d_2, d_3) using the parameters (d_1, m, Δ), where $\Delta = d_1 + d_2 + d_3$. In terms of this data we have $d_2 = m - d_1$ and $d_3 = \Delta - m$. We also introduce the notation

(7.1)
$$F(\alpha, \varpi) = \sum_{p \in D} (\alpha_1(p) + \alpha_2(p) + \alpha_3(p) - 3\alpha_{\varpi(3)}(p)),$$
$$G(\alpha, \varpi) = \sum_{p \in D} (2\alpha_1(p) + 2\alpha_2(p) + 2\alpha_3(p) - 3\alpha_{\varpi(2)}(p) - 3\alpha_{\varpi(3)}(p)).$$

PROPOSITION 7.1. *A parabolic Higgs bundle $(L_1 \oplus L_2 \oplus L_3, \Phi)$ of type $(1,1,1)$ is stable if and only if the maps Φ_1 and Φ_2 are non-zero and, furthermore,*

$$3m > 2\Delta - F(\alpha, \varpi),$$
$$3d_1 > \Delta - G(\alpha, \varpi).$$

PROOF. It is immediate from Propositions 3.4 and 3.6 that a parabolic Higgs bundle of type $(1,1,1)$ is stable if and only if the conditions $\mathrm{par}\mu(L_3) < \mathrm{par}\mu(E)$ and $\mathrm{par}\mu(L_2 \oplus L_3) < \mathrm{par}\mu(E)$ hold and $\Phi_l \neq 0$ for $l = 1, 2$. From this we obtain the characterization given in the proposition by calculating the relevant parabolic

degrees:
$$\operatorname{pardeg}(E) = \Delta + \sum_{p \in D} (\alpha_1(p) + \alpha_2(p) + \alpha_3(p)) ,$$
$$\operatorname{pardeg}(L_3) = d_3 + \sum_{p \in D} \alpha_{\varpi(3)}(p)$$
$$= \Delta - m + \sum_{p \in D} \alpha_{\varpi(3)}(p) ,$$
$$\operatorname{pardeg}(L_2 \oplus L_3) = d_2 + d_3 + \sum_{p \in D} (\alpha_{\varpi(2)}(p) + \alpha_{\varpi(3)}(p))$$
$$= \Delta - d_1 + \sum_{p \in D} (\alpha_{\varpi(2)}(p) + \alpha_{\varpi(3)}(p)) .$$

Thus we get
$$\operatorname{par}\mu(L_3) < \operatorname{par}\mu(E)$$
$$\iff 3\Delta - 3m + 3\sum_{p \in D} \alpha_{\varpi(3)}(p) < \Delta + \sum_{p \in D} (\alpha_1(p) + \alpha_2(p) + \alpha_3(p))$$
$$\iff 3m > 2\Delta - F(\alpha, \varpi) ,$$

while
$$\operatorname{par}\mu(L_2 \oplus L_3) < \operatorname{par}\mu(E)$$
$$\Leftrightarrow \quad 3\Delta - 3d_1 + 3 \sum_{p \in D} (\alpha_{\varpi(2)}(p) + \alpha_{\varpi(3)}(p)) < 2\Delta + 2 \sum_{p \in D} (\alpha_1(p) + \alpha_2(p) + \alpha_3(p))$$
$$\Leftrightarrow \quad 3d_1 > \Delta - G(\alpha, \varpi) .$$
$\square$

Denote by $\mathcal{N}_{(1,1,1)}(d_1, m, \varpi)$ the critical submanifold of parabolic Higgs bundles of type $(1,1,1)$ with invariants (d_1, m) and weights given by $\varpi = \{\varpi_p\}_{p \in D}$. Introduce the following notation

(7.2)
$$\begin{aligned} S_1 &= \{p \in D \mid \alpha_{\varpi(1)}(p) > \alpha_{\varpi(2)}(p)\}, \\ S_2 &= \{p \in D \mid \alpha_{\varpi(2)}(p) > \alpha_{\varpi(3)}(p)\}, \\ s_1 &= \#S_1, \ s_2 = \#S_2 . \end{aligned}$$

There is no risk of confusion with the sets S_1, S_2 defined in Chapter 6, since this notation will only apply to this chapter. By Lemma 6.1,
$$\operatorname{SParHom}(L_l, L_{l+1} \otimes K(D)) \cong \operatorname{Hom}(L_l, L_{l+1} \otimes K(D - S_l))$$
for $l = 1, 2$. We let for $l = 1, 2$,
$$m_l = \deg(\operatorname{Hom}(L_l, L_{l+1} \otimes K(D - S_l))).$$
Then

(7.3)
$$\begin{aligned} m_1 &= d_2 - d_1 + 2g - 2 + n - s_1 \\ &= m - 2d_1 + n - s_1 + 2g - 2, \\ m_2 &= d_3 - d_2 + 2g - 2 + n - s_2 \\ &= \Delta - 2m + d_1 + n - s_2 + 2g - 2. \end{aligned}$$

1. DESCRIPTION OF THE CRITICAL SUBMANIFOLDS

PROPOSITION 7.2. *The critical submanifold $\mathcal{N}_{(1,1,1)}(d_1, m, \varpi)$ is non-empty if and only if the following conditions are satisfied*

$$\begin{array}{rl} & 3m > 2\Delta - F(\alpha, \varpi), \\ & 3d_1 > \Delta - G(\alpha, \varpi), \\ m_1 \geq 0 \iff & 2d_1 - m \leq n - s_1 + 2g - 2, \\ m_2 \geq 0 \iff & 2m - d_1 \leq \Delta + n - s_2 + 2g - 2, \end{array}$$

where F and G were defined in (7.1). Moreover, the map

$$\mathcal{N}_{(1,1,1)}(d_1, m, \varpi) \to \mathrm{Jac}^{d_1}(X) \times S^{m_1}X \times S^{m_2}X$$
$$(L_1 \oplus L_2 \oplus L_3, \Phi_1, \Phi_2) \mapsto (L_1, \mathrm{div}(\Phi_1), \mathrm{div}(\Phi_2))$$

is an isomorphism.

PROOF. Proposition 7.1 shows that the conditions in the statement are necessary for $\mathcal{N}_{(1,1,1)}(d_1, m, \varpi)$ to be non-empty.

Assume now that we are given (d_1, m, ϖ) satisfying these conditions. For any line bundle L_1 in $\mathrm{Jac}^{d_1}(X)$ and effective divisors $D_1 \in S^{m_1}X$ and $D_2 \in S^{m_2}X$ we get line bundles $M_l = \mathcal{O}(D_l)$ with non-zero sections Φ_l determined up to multiplication by nonzero scalars for $l = 1, 2$. We then obtain a parabolic Higgs bundle of type $(1, 1, 1)$ by letting

$$L_2 = L_1 \otimes K^{-1}(S_1 - D) \otimes M_1,$$
$$L_3 = L_2 \otimes K^{-1}(S_2 - D) \otimes M_2,$$

and defining Φ to have components Φ_1 and Φ_2. Clearly this parabolic Higgs bundle has the desired invariants (d_1, m, ϖ) and, if the conditions in the statement are satisfied, then Proposition 7.1 shows that it is indeed stable.

It follows from this construction that the map given in the statement of the proposition is surjective. To see that it is injective, we note that taking non-zero scalar multiples of the Higgs fields $\Phi_1 \in H^0(L_1^{-1} \otimes L_2 \otimes K(D - S_1))$ and $\Phi_2 \in H^0(L_2^{-1} \otimes L_3 \otimes K(D - S_2))$ gives rise to isomorphic parabolic Higgs bundles of type $(1, 1, 1)$. Thus the map given is, in fact, an isomorphism. □

COROLLARY 7.3. *The critical submanifold $\mathcal{N}_{(1,1,1)}(d_1, m, \varpi)$ has Poincaré polynomial $P_t(\mathcal{N}_{(1,1,1)}(d_1, m, \varpi))$ equal to*

$$(1+t)^{2g} \underset{x^0 y^0}{\mathrm{Coeff}} \left(\frac{(1+xt)^{2g}}{(1-x)(1-xt^2)x^{m_1}} \cdot \frac{(1+yt)^{2g}}{(1-y)(1-yt^2)y^{m_2}} \right).$$

PROOF. Immediate from MacDonald's formula [29] for the Poincaré polynomial of a symmetric product of X. □

The total contribution to the Poincaré polynomial of $\mathcal{M}$ from submanifolds of type $(1, 1, 1)$ is

$$(7.4) \qquad P_t(\Delta, (1,1,1)) := \sum_{d_1, m, \varpi} t^{\lambda_{(d_1, m, \varpi)}} P_t(\mathcal{N}_{(1,1,1)}(d_1, m, \varpi)),$$

where $\lambda_{(d_1, m, \varpi)}$ is the index of the critical submanifold $\mathcal{N}_{(1,1,1)}(d_1, m, \varpi)$ and the sum is over all permutations $\varpi = \{\varpi_p\}_{p \in D} \in (S_3)^n$ and pairs of integers (d_1, m) such that the bounds of Proposition 7.2 are satisfied.

7. CRITICAL SUBMANIFOLDS OF TYPE (1, 1, 1)

LEMMA 7.4. *The index of the critical submanifold* $\mathcal{N}_{(1,1,1)}(d_1, m, \varpi)$ *is*
$$\lambda_{(d_1, m, \varpi)} = 2(4g - 4 + n + s_1 + s_2 - \Delta + d_1 + m).$$

PROOF. The formula for the Morse index is given in Proposition 3.11.

We need to calculate, for each $p \in D$ the numbers f_p and the dimensions of the spaces P_p and N_p, which enter in this formula.

Recall (from Proposition 2.4) that $f_p = (1/2)(r^2 - \sum_i m_i(p)^2)$. Since the multiplicities are all 1 and $r = 3$ we get $f_p = (1/2)(9 - (1 + 1 + 1)) = 3$ and hence
$$\sum_p f_p = 3n \ .$$

For $p \in D$, the space $P_p(L_l, L_l)$ consists of the parabolic endomorphisms of $L_{l,p}$, so $\dim P_p(L_l, L_l) = 1$. The space $N_p(L_l, L_{l+1})$ is the space of strictly parabolic maps from $L_{l,p}$ to $L_{l+1,p}$ and hence
$$N_p(L_l, L_{l+1}) = \begin{cases} 0 & \text{if } \alpha_{\varpi(l)}(p) > \alpha_{\varpi(l+1)}(p), \\ \text{Hom}(L_{l,p}, L_{l+1,p}) & \text{otherwise.} \end{cases}$$

Recalling from (7.2) the definition of S_l, it follows that
$$\dim N_p(L_l, L_{l+1}) = \begin{cases} 0 & \text{if } p \in S_l, \\ 1 & \text{if } p \in D - S_l, \end{cases}$$

and thus
$$\sum_p \dim N_p(L_l, L_{l+1}) = n - s_l \ .$$

Substituting this in the formula for the Morse index we get
$$\lambda_{(d_1, m, \varpi)} = r^2(2g - 2) + 2\sum_p f_p + 2\sum_{l=1}^{3}\left((1 - g - n)\operatorname{rk}(L_l)^2 + \sum_p \dim P_p(L_l, L_l)\right)$$
$$+ 2\sum_{l=1}^{2}\left((1 - g)\operatorname{rk}(L_l)\operatorname{rk}(L_{l+1}) - \operatorname{rk}(L_l)\deg(L_{l+1}) + \operatorname{rk}(L_{l+1})\deg(L_l)\right.$$
$$\left. - \sum_p \dim N_p(L_l, L_{l+1})\right)$$
$$= 9(2g - 2) + 2 \cdot 3n + 2(3(1 - g - n) + 3n)$$
$$+ 2(2(1 - g) - d_2 - d_3 + d_1 + d_2 - (n - s_1 + n - s_2))$$
$$= 2(4g - 4 + n + s_1 + s_2 - \Delta + d_1 + m) \ .$$
$\square$

2. The sum for fixed ϖ.

We shall now calculate the total contribution (7.4) to the Poincaré polynomial of $\mathcal{M}$ from submanifolds of type $(1, 1, 1)$ in several stages. We begin by doing the sum over (d_1, m) for a fixed permutation ϖ.

LEMMA 7.5. *Let* $\varpi = \{\varpi_p\}_{p \in D} \in (S_3)^n$ *be fixed. Then*
$$\sum_{d_1, m} t^{\lambda_{(d_1, m, \varpi)}} P_t(\mathcal{N}_{(1,1,1)}(d_1, m, \varpi)) = \operatorname*{Coeff}_{x^0 y^0} \Psi(\varpi),$$

where we have defined

$$\Psi(\varpi) =$$
$$\sum_{\substack{d_1 \geq \bar{d}_1 \\ m \geq \bar{m}}} t^{2(4g-4+n+s_1+s_2-\Delta+d_1+m)}(1+t)^{2g}\frac{(1+xt)^{2g}}{(1-x)(1-xt^2)x^{m_1}} \cdot \frac{(1+yt)^{2g}}{(1-y)(1-yt^2)y^{m_2}}$$

with

(7.5)
$$\bar{m} = [(2/3)\Delta - (1/3)F(\alpha,\varpi) + 1],$$
$$\bar{d}_1 = [(1/3)\Delta - (1/3)G(\alpha,\varpi) + 1].$$

PROOF. The identity would be clear from Corollary 7.3 and Lemma 7.4 if the latter sum were over (d_1, m) satisfying the conditions of Proposition 7.2. Now, from these equations we see that we need to sum over the closed region in the (m, d_1)-plane bounded by the lines

$$m = \bar{m},$$
$$d_1 = \bar{d}_1,$$
$$2d_1 - m = n - s_1 + 2g - 2,$$
$$2m - d_1 = \Delta + n - s_2 + 2g - 2.$$

Thus, summing over the semi-infinite region defined by $d_1 \geq \bar{d}_1$ and $m \geq \bar{m}$, we introduce in the sum extra terms. But, since the lines given by the third and fourth equations represent the conditions $m_1 \geq 0$ and $m_2 \geq 0$, these extra terms have strictly positive powers of x or y and hence this does not change the coefficient of $x^0 y^0$. $\square$

Using (7.3) we have that

$$x^{m_1} = x^{n-s_1+2g-2}x^{-2d_1}x^m \quad \text{and} \quad y^{m_2} = y^{\Delta+n-s_2+2g-2}y^{d_1}y^{-2m},$$

hence

$$\Psi(\varpi) = t^{2(4g-4+n+s_1+s_2-\Delta)}(1+t)^{2g}\frac{(1+xt)^{2g}}{(1-x)(1-xt^2)} \cdot \frac{(1+yt)^{2g}}{(1-y)(1-yt^2)}$$
$$\cdot \frac{1}{x^{n-s_1+2g-2}y^{\Delta+n-s_2+2g-2}} \sum_{d_1=\bar{d}_1}^{\infty} \frac{t^{2d_1}x^{2d_1}}{y^{d_1}} \sum_{m=\bar{m}}^{\infty} \frac{t^{2m}y^{2m}}{x^m}$$

(7.6)
$$= t^{2(4g-4+n+s_1+s_2-\Delta)}(1+t)^{2g}\frac{(1+xt)^{2g}}{(1-x)(1-xt^2)} \cdot \frac{(1+yt)^{2g}}{(1-y)(1-yt^2)}$$
$$\cdot \frac{(x^2 y^{-1} t^2)^{\bar{d}_1}(x^{-1}y^2 t^2)^{\bar{m}}}{x^{n-s_1+2g-2}y^{\Delta+n-s_2+2g-2}(1-x^2 y^{-1}t^2)(1-x^{-1}y^2 t^2)}$$
$$= t^{2(s_1+s_2)}x^{s_1}y^{s_2}(x^2 y^{-1}t^2)^{\bar{d}_1}(x^{-1}y^2 t^2)^{\bar{m}} \cdot \frac{t^{2(4g-4+n-\Delta)}}{x^{n+2g-2}y^{\Delta+n+2g-2}}$$
$$\cdot \frac{(1+t)^{2g}(1+xt)^{2g}(1+yt)^{2g}}{(1-x)(1-xt^2)(1-y)(1-yt^2)(1-x^2 y^{-1}t^2)(1-x^{-1}y^2 t^2)},$$

where we have separated powers of x, y and t which potentially depend on ϖ.

3. The sum over ϖ.

In order to proceed with the calculation we need to sum the contribution (7.6) over all permutations $\varpi = (\varpi_p) \in (S_3)^n$. For this we need to understand the dependence of s_1, s_2, $\bar{m}$ and $\bar{d}_1$ on ϖ. Now, looking at the definitions (7.5), we see that $\bar{d}_1$ and $\bar{m}$ also depend on the weights. In order to deal with this dependence, we shall take advantage of Proposition 2.1 which allows us to do the computation in the case where the degree Δ satisfies that $\Delta \not\equiv 0 \pmod{3}$. Hence we can choose the weights so as to facilitate the computations, as long as we keep the same choice throughout. Now, if $\Delta \not\equiv 0 \pmod{3}$ and we choose the weights $\alpha_i(p)$ sufficiently small, then $\bar{d}_1$ and $\bar{m}$ are independent of ϖ. For future reference, we state here our assumptions.

ASSUMPTION 7.6. Write $D = p_1 + \cdots + p_n$. In addition to Assumption 5.1, we shall from now on assume that $\Delta \not\equiv 0 \pmod{3}$ and that the weights $\alpha_i(p)$ are chosen to satisfy

$$\alpha_i(p_j) \ll 1 \quad \text{for all } i, j.$$

Next we consider the dependence of s_1 and s_2 on ϖ. We can write

$$s_1 = \sum_{p \in D} s_1(p),$$

$$s_2 = \sum_{p \in D} s_2(p),$$

where $s_1(p)$ and $s_2(p)$ are defined in the obvious way:

$$s_1(p) = \begin{cases} 1 & \text{if } \varpi_p(1) > \varpi_p(2), \\ 0 & \text{otherwise}, \end{cases}$$

and

$$s_2(p) = \begin{cases} 1 & \text{if } \varpi_p(2) > \varpi_p(3), \\ 0 & \text{otherwise}. \end{cases}$$

We give the values of $s_1(p)$, $s_2(p)$ and $s_1(p) + s_2(p)$ as a function of ϖ_p in Table 7.1, using the notation $\varpi = (\varpi(1)\,\varpi(2)\,\varpi(3))$ for a permutation $\varpi \in S_3$.

TABLE 7.1. $s_1(p)$ and $s_2(p)$ as a function of ϖ_p

ϖ_p	(123)	(231)	(312)	(213)	(132)	(321)
$s_1(p)$	0	0	1	1	0	1
$s_2(p)$	0	1	0	0	1	1
$s_1(p) + s_2(p)$	0	1	1	1	1	2

Under Assumption 7.6, $\bar{m}$ and $\bar{d}_1$ are independent of ϖ: in fact we have from the definitions (7.5) of $\bar{m}$ and $\bar{d}_1$ that

(7.7)
$$\bar{m} = \left[\tfrac{2\Delta}{3}\right] + 1,$$
$$\bar{d}_1 = \left[\tfrac{\Delta}{3}\right] + 1.$$

Therefore to do the sum $\sum_\varpi \Psi(\varpi)$, we only need to do $\sum_{\varpi\in(S_3)^n} t^{2(s_1+s_2)}x^{s_1}y^{s_2}$ in (7.6). For this we use Table 7.1 and obtain:

(7.8)
$$\sum_{\varpi\in(S_3)^n} t^{2(s_1+s_2)}x^{s_1}y^{s_2} = \prod_{p\in D}\sum_{\varpi_p\in S_3} t^{2(s_1(p)+s_2(p))}x^{s_1(p)}y^{s_2(p)}$$
$$= \prod_{p\in D}(1+2t^2x+2t^2y+t^4xy)$$
$$= (1+2t^2x+2t^2y+t^4xy)^n.$$

Combining (7.8) with (7.6) we finally obtain:

(7.9)
$$\sum_\varpi \Psi(\varpi) = (1+2t^2x+2t^2y+t^4xy)^n \cdot$$
$$\cdot \frac{t^{2(4g-4+n-\Delta)}(1+t)^{2g}(1+xt)^{2g}(1+yt)^{2g}(x^2y^{-1}t^2)^{\bar{d}_1}(x^{-1}y^2t^2)^{\bar{m}}}{x^{n+2g-2}y^{\Delta+n+2g-2}(1-x)(1-xt^2)(1-y)(1-yt^2)(1-x^2y^{-1}t^2)(1-x^{-1}y^2t^2)}.$$

Note that this expression has arbitrarily large positive and negative powers of x and y. Therefore it is not very suitable for extracting the coefficient to x^0y^0. However, to facilitate this task we can make the following change of variable:
$$x = u^2v, \qquad y = uv^2.$$

Then we have
$$x^2y^{-1} = u^3, \qquad x^{-1}y^2 = v^3 \quad \text{and} \quad xy = u^3v^3.$$

Substituting in (7.9), and using (7.7), we finally obtain the formula for the contribution to the Poincaré polynomial from critical submanifolds of type $(1,1,1)$:

PROPOSITION 7.7. *Under Assumption 7.6, let $\Delta_0 \in \{1,2\}$ be the remainder modulo 3 of Δ. Then*

$$P_t(\Delta,(1,1,1)) = \operatorname*{Coeff}_{u^0v^0}\Bigg((1+2u^2vt^2+2uv^2t^2+u^3v^3t^4)^n \cdot$$
$$\cdot \frac{t^{2(4g-3+n)}}{u^{3n+6g-9+\Delta_0}v^{3n+6g-6-\Delta_0}} \cdot$$
$$\cdot \frac{(1+t)^{2g}(1+u^2vt)^{2g}(1+uv^2t)^{2g}}{(1-u^2v)(1-uv^2)(1-u^2vt^2)(1-uv^2t^2)(1-v^3t^2)(1-u^3t^2)}\Bigg).$$

□

CHAPTER 8

Critical submanifolds of type $(1,2)$

1. Description of the critical submanifolds

In this chapter, we consider the critical points of the Bott–Morse function f represented by Higgs bundles (E, Φ) which are of the form $E = E_0 \oplus E_1$ where $E_0 = L$ is a parabolic line bundle, E_1 is a rank 2 parabolic bundle and $\Phi : L \to E_1 \otimes K(D)$ is a strongly parabolic homomorphism. This defines a parabolic triple $(E_1 \otimes K, L)$ of type $(1,2)$. By Proposition 4.6, the triple is σ-stable exactly for the value $\sigma = 2g - 2$.

In order to do the computations, let us introduce some notation. Recall that we keep our assumption of generic weights. The (fixed) weights of E at each $p \in D$ are $0 < \alpha_1(p) < \alpha_2(p) < \alpha_3(p) < 1$. Each possible choice of distribution of these weights is given by a function $\varpi = \{\varpi_p\}_{p \in D}$ that assigns to each $p \in D$ a number $\varpi(p) \in \{1, 2, 3\}$ such that the weight of L is $\alpha(p) = \alpha_{\varpi(p)}(p)$. The weights of E_1 are $\beta_1(p) < \beta_2(p)$, so that $\{\beta_1(p), \beta_2(p), \alpha(p)\} = \{\alpha_1(p), \alpha_2(p), \alpha_3(p)\}$. In the decomposition $E = L \oplus E_1$, we define

$$d_1 = \deg(E_1 \otimes K) = \deg(E_1) + 4g - 4,$$
$$d_2 = \deg(L),$$

so that $\Delta = \deg(E) = d_1 + d_2 + 4 - 4g$.

Denote by $\mathcal{N}_{(1,2)}(d_1, \varpi)$ the critical submanifold of parabolic Higgs bundles of type $(1,2)$ with topological invariants given by $(d_1, d_2 = \Delta - d_1 + 4g - 4)$ and weights determined by ϖ. The contribution of all critical submanifolds of type $(1,2)$ is given as

$$P_t(\Delta, (1,2)) := \sum_{d_1, \varpi} t^{\lambda_{(d_1, \varpi)}} P_t(\mathcal{N}_{(1,2)}(d_1, \varpi)),$$

where $\lambda_{(d_1, \varpi)}$ is the index of $\mathcal{N}_{(1,2)}(d_1, \varpi)$.

LEMMA 8.1. *The index of the critical submanifold $\mathcal{N}_{(1,2)}(d_1, \varpi)$ is*

$$\lambda_{(d_1, \varpi)} = 12g - 12 + 4n - 2d_1 + 4d_2 - 2s_0,$$

where $s_0 = \#\{\beta_i(p) \,|\, \beta_i(p) > \alpha(p)\}$.

PROOF. From Proposition 2.4, $f_p = \frac{1}{2}(r^2 - \sum_i m_i(p)^2) = 3$, since the multiplicities are all 1. For $p \in D$, the space $P_p(L, L)$ consists of endomorphisms of L_p, so $\dim P_p(L, L) = 1$ and $P_p(E_1 \otimes K, E_1 \otimes K)$ consists of parabolic endomorphisms of $(E_1 \otimes K)_p$, so it has dimension 3. The dimension of the space of strongly parabolic homomorphisms from L_p to $(E_1 \otimes K)_p$ is given by

$$\dim N_p(L, E_1 \otimes K) = \begin{cases} 2 & \text{if } \alpha(p) < \beta_1(p), \\ 1 & \text{if } \beta_1(p) < \alpha(p) < \beta_2(p), \\ 0 & \text{if } \beta_2(p) < \alpha(p). \end{cases}$$

Therefore
$$s_0 = \sum_p \dim N_p(L, E_1 \otimes K) = \#\{\beta_i(p) \mid \beta_i(p) > \alpha(p)\}.$$

Substituting this in the formula for the Morse index in Proposition 3.11, we have

$$\lambda_{(d,\varpi)} =$$
$$= 9(2g-2) + 6n + 2\big(5(1-g-n) + 2(1-g) - (d_1 - 4g + 4) + 2d_2 + 4n - s_0\big)$$
$$= 12g - 12 + 4n - 2d_1 + 4d_2 - 2s_0.$$

$\square$

By Proposition 4.6, $\mathcal{N}_{(1,2)}(d_1, \varpi)$ is isomorphic to the moduli space of σ-stable triples of the appropriate type with $\sigma = 2g - 2$. By the genericity of weights, such σ is not a critical value. Its Poincaré polynomial is given by Theorem 6.4. So, for each $\varepsilon = \{\varepsilon(p)\}_{p \in D}$, let s_1, s_2, s_3 be defined by (6.2), $\varsigma(p) = 3 - \varepsilon(p)$ and

$$\bar{d}_M = \left[\frac{1}{3}\left(d_1 + d_2 + \sum \big(\alpha(p) + \beta_{\varsigma(p)}(p) - 2\beta_{\varepsilon(p)}(p)\big) + 2g - 2\right)\right] + 1$$
$$= \left[\frac{1}{3}\left(\Delta + \sum \big(\alpha(p) + \beta_{\varsigma(p)}(p) - 2\beta_{\varepsilon(p)}(p)\big)\right)\right] + 2g - 1.$$

Then Theorem 6.4, Lemma 8.1 and the fact that $s_0 = s_1 + s_2$ yield that

(8.1)
$$P_t(\Delta, (1,2)) = \sum_{d_1, \varpi} t^{12g-12+4n-2d_1+4d_2-2s_1-2s_2}.$$

$$\sum_{\varepsilon} \operatorname*{Coeff}_{x^0}\left(\frac{(1+t)^{4g}(1+xt)^{2g}t^{2d_1-2d_2+2s_2+2s_3-2\bar{d}_M}x^{\bar{d}_M}}{(1-t^2)(1-x)(1-xt^2)(1-t^{-2}x)x^{d_1-d_2+s_1}}\right.$$
$$\left. - \frac{(1+t)^{4g}(1+xt)^{2g}t^{-2d_1+2g-2+2n-2s_3+4\bar{d}_M}x^{\bar{d}_M}}{(1-t^2)(1-x)(1-xt^2)(1-t^4x)x^{d_1-d_2+s_1}}\right).$$

2. The sum for fixed (ϖ, ε).

We shall compute the contribution (8.1) to the Poincaré polynomial of $\mathcal{M}$ from submanifolds of type $(1,2)$. We start by performing the sum over all possibilities of d_1 for each choice of (ϖ, ε). The condition that the moduli space $\mathcal{N}_{(1,2)}(d_1, \varpi)$ be non-empty is $2g - 2 > \operatorname{par}\mu_1 - \operatorname{par}\mu_2$ (see Remark 6.5). This means that

$$2g - 2 > d_1/2 - d_2 + \sum \big(\beta_1(p) + \beta_2(p) - 2\alpha(p)\big)/2.$$

Using that $\Delta = d_1 + d_2 + 4 - 4g$, this is translated into

$$d_2 - d_1 > 4 - 4g - \frac{\Delta}{3} + \frac{2}{3}\sum \big(\beta_1(p) + \beta_2(p) - 2\alpha(p)\big).$$

But $d_2 - d_1 \equiv \Delta \pmod{2}$. So $d_2 - d_1 = \bar{d}_0 + 2k$, $k \geq 0$, and

$$\bar{d}_0 = 4 - 4g + 2\left[\frac{1}{2}\left(\left[-\frac{\Delta}{3} + \frac{2}{3}\sum \big(\beta_1(p) + \beta_2(p) - 2\alpha(p)\big)\right] + \Delta\right)\right] - \Delta + 2.$$

This gives the range for the summation in (8.1) for d_1 for fixed (ϖ, ε). Looking at (8.1), one sees that we need to compute

$$\sum t^{2d_2} x^{d_2-d_1} = t^{\Delta+4g-4} \sum t^{d_2-d_1} x^{d_2-d_1} = \frac{t^{\bar{d}_0+\Delta+4g-4} x^{\bar{d}_0}}{1-t^2x^2},$$

$$\sum t^{-4d_1+4d_2} x^{d_2-d_1} = \frac{t^{4\bar{d}_0} x^{\bar{d}_0}}{1-t^8x^2}.$$

Substituting into (8.1), we get that

(8.2)

$$P_t(\Delta, (1,2)) =$$
$$= \sum_{\varpi,\varepsilon} \operatorname*{Coeff}_{x^0} \left(\frac{(1+t)^{4g}(1+xt)^{2g} t^{16g-16+4n-2s_1+2s_3-2\bar{d}_M+\bar{d}_0+\Delta} x^{\bar{d}_M+\bar{d}_0-s_1}}{(1-t^2)(1-x)(1-xt^2)(1-t^{-2}x)(1-t^2x^2)} \right.$$
$$\left. - \frac{(1+t)^{4g}(1+xt)^{2g} t^{14g-14+6n-2s_1-2s_2-2s_3+4\bar{d}_M+4\bar{d}_0} x^{\bar{d}_M+\bar{d}_0-s_1}}{(1-t^2)(1-x)(1-xt^2)(1-t^4x)(1-t^8x^2)} \right).$$

3. The sum over ϖ and ε.

Now we need to perform the sum in (8.2) for all choices of (ϖ, ε). For this we arrange the degree and the weights to satisfy Assumption 7.6. Write $\Delta = 3k + \Delta_0$, $\Delta_0 \in \{1, 2\}$. Since $\alpha_i(p)$ are sufficiently small, we have that

(8.3)
$$\bar{d}_M = \left[\frac{\Delta}{3}\right] + 2g - 1 = k + 2g - 1,$$
$$\bar{d}_0 = 4 - 4g + 2\left[\frac{1}{2}\left(\left[-\frac{\Delta}{3}\right] + \Delta\right)\right] - \Delta + 2 = 6 - 4g - k - \Delta_0$$

are independent of (ϖ, ε). Therefore to do the sum (8.2), we only need to do

$$\sum_{\varpi,\varepsilon} t^{-2s_1+2s_3} x^{-s_1} \quad \text{and} \quad \sum_{\varpi,\varepsilon} t^{-2s_1-2s_2-2s_3} x^{-s_1}.$$

We have to write down the dependence of s_1, s_2, s_3 on (ϖ, ε). Note that we can write

$$s_1 = \sum_{p \in D} s_1(p),$$
$$s_2 = \sum_{p \in D} s_2(p),$$
$$s_3 = \sum_{p \in D} s_3(p),$$

where $s_1(p)$, $s_2(p)$ and $s_3(p)$ are defined in the obvious way:

$$s_1(p) = \begin{cases} 1 & \text{if } \alpha(p) < \beta_{\varsigma(p)}(p), \\ 0 & \text{otherwise}, \end{cases}$$

$$s_2(p) = \begin{cases} 1 & \text{if } \alpha(p) < \beta_{\varepsilon(p)}(p), \\ 0 & \text{otherwise}, \end{cases}$$

and
$$s_3(p) = \begin{cases} 1 & \text{if } \beta_{\varepsilon(p)}(p) < \beta_{\varsigma(p)}(p), \\ 0 & \text{otherwise.} \end{cases}$$

We give the values of $s_1(p)$, $s_2(p)$ and $s_3(p)$ as a function of $(\varpi(p), \varepsilon(p)) \in \{1, 2, 3\} \times \{1, 2\}$ in Table 8.1,

TABLE 8.1. $s_1(p)$, $s_2(p)$ and $s_3(p)$ as a function of $(\varpi(p), \varepsilon(p))$

$(\varpi(p), \varepsilon(p))$	$(1,1)$	$(1,2)$	$(2,1)$	$(2,2)$	$(3,1)$	$(3,2)$
$s_1(p)$	1	1	1	0	0	0
$s_2(p)$	1	1	0	1	0	0
$s_3(p)$	1	0	1	0	1	0

We obtain
$$\sum_{\varpi,\varepsilon} t^{-2s_1+2s_3} x^{-s_1} = \prod_{p \in D} \sum_{\varpi(p),\varepsilon(p)} t^{-2s_1(p)+2s_3(p)} x^{-s_1(p)}$$
$$= \prod_{p \in D} (t^{-2}x^{-1} + 2 + 2x^{-1} + t^2)$$
$$= (t^{-2}x^{-1} + 2 + 2x^{-1} + t^2)^n$$
$$= t^{-2n} x^{-n} (1 + 2t^2 + 2t^2 x + t^4 x)^n,$$

and
$$\sum_{\varpi,\varepsilon} t^{-2s_1-2s_2-2s_3} x^{-s_1} = \prod_{p \in D} \sum_{\varpi(p),\varepsilon(p)} t^{-2s_1(p)-2s_2(p)-2s_3(p)} x^{-s_1(p)}$$
$$= (2t^{-4}x^{-1} + x^{-1}t^{-6} + 1 + 2t^{-2})^n$$
$$= t^{-6n} x^{-n} (1 + 2t^2 + 2t^4 x + t^6 x)^n.$$

Combining this with (8.2) we get that $P_t(\Delta, (1,2))$ equals
$$\operatorname*{Coeff}_{x^0} \left(\frac{(1+t)^{4g}(1+xt)^{2g} t^{16g-16+2n-2\bar{d}_M+\bar{d}_0+\Delta} x^{\bar{d}_M+\bar{d}_0-n}(1+2t^2+2t^2x+t^4x)^n}{(1-t^2)(1-x)(1-xt^2)(1-t^{-2}x)(1-t^2x^2)} \right.$$
$$\left. - \frac{(1+t)^{4g}(1+xt)^{2g} t^{14g-14+4\bar{d}_M+4\bar{d}_0} x^{\bar{d}_M+\bar{d}_0-n}(1+2t^2+2t^4x+t^6x)^n}{(1-t^2)(1-x)(1-xt^2)(1-t^4x)(1-t^8x^2)} \right).$$

Now, using that $\bar{d}_M + \bar{d}_0 = 5 - 2g - \Delta_0$ and $-3\bar{d}_M + \Delta = 3 - 6g + \Delta_0$, which follow from (8.3), we have the following.

PROPOSITION 8.2. *Under Assumption 7.6, let $\Delta_0 \in \{1, 2\}$ be the remainder modulo 3 of Δ. Then $P_t(\Delta, (1,2))$ equals*
$$\operatorname*{Coeff}_{x^0} \left(\frac{(1+t)^{4g}(1+xt)^{2g} t^{8g-8+2n} x^{5-2g-\Delta_0-n}(1+2t^2+2t^2x+t^4x)^n}{(1-t^2)(1-x)(1-xt^2)(1-t^{-2}x)(1-t^2x^2)} \right.$$
$$\left. - \frac{(1+t)^{4g}(1+xt)^{2g} t^{6g+6-4\Delta_0} x^{5-2g-\Delta_0-n}(1+2t^2+2t^4x+t^6x)^n}{(1-t^2)(1-x)(1-xt^2)(1-t^4x)(1-t^8x^2)} \right).$$
□

CHAPTER 9

Critical submanifolds of type $(2,1)$

1. Description of the critical submanifolds

In this chapter, we consider the critical points of the Bott–Morse function f represented by Higgs bundles (E, Φ) which are of the form $E = E_0 \oplus E_1$ where $E_1 = L$ is a parabolic line bundle, E_0 is a rank 2 parabolic bundle and $\Phi : E_0 \to L \otimes K(D)$ is a strongly parabolic homomorphism. This defines a parabolic triple $(L \otimes K, E_0)$ of type $(2,1)$. By Proposition 4.6, the triple is σ-stable exactly for the value $\sigma = 2g - 2$.

As in Chapter 8, the (fixed) weights of E at each $p \in D$ are $0 < \alpha_1(p) < \alpha_2(p) < \alpha_3(p) < 1$. Each possible choice of distribution of these weights is given by some $\varpi = \{\varpi(p)\}_{p \in D}$ where $\varpi(p) \in \{1, 2, 3\}$, $p \in D$ such that the weight of L is $\alpha(p) = \alpha_{\varpi(p)}(p)$. The weights of E_0 are $\beta_1(p) < \beta_2(p)$, so that $\{\beta_1(p), \beta_2(p), \alpha(p)\} = \{\alpha_1(p), \alpha_2(p), \alpha_3(p)\}$. In the decomposition $E = E_0 \oplus L$, we define

$$\begin{aligned} d_1 &= \deg(L \otimes K) = \deg(L) + 2g - 2, \\ d_2 &= \deg(E_0), \end{aligned}$$

so that $\Delta = \deg(E) = d_1 + d_2 + 2 - 2g$.

Denote by $\mathcal{N}_{(2,1)}(d_1, \varpi)$ the critical submanifold of parabolic Higgs bundles of type $(2,1)$ with topological invariants given by $(d_1, d_2 = \Delta - d_1 + 2g - 2)$ and where the weights are determined by ϖ.

LEMMA 9.1. $\mathcal{N}_{(2,1)}(d_1, \varpi)$ *is isomorphic to the moduli space of σ-stable parabolic triples of type $(1,2)$ with degrees $d_2' = -n - d_1, d_1' = -2n - d_2$ and weights $1 - \alpha(p)$ for the line bundle and $1 - \beta_2(p) < 1 - \beta_1(p)$ for the rank 2-bundle, for $\sigma = 2g - 2$.*

PROOF. The result follows by dualizing and applying Proposition 4.2. And by the definition of dual of a parabolic bundle. □

Note that by the genericity of weights, the value $\sigma = 2g - 2$ is not a critical value for the moduli space of parabolic triples. Now let $\varepsilon = \{\varepsilon(p)\}_{p \in D}$, $\varepsilon(p) \in \{1, 2\}$, and $\varsigma(p) = 3 - \varepsilon(p)$. We introduce the following sets:

$$\begin{aligned} S_1 &= \{p \in D \mid 1 - \alpha(p) < 1 - \beta_{\varsigma(p)}(p)\} = \{p \in D \mid \alpha(p) > \beta_{\varsigma(p)}(p)\}, \\ S_2 &= \{p \in D \mid 1 - \alpha(p) < 1 - \beta_{\varepsilon(p)}(p)\} = \{p \in D \mid \alpha(p) > \beta_{\varepsilon(p)}(p)\}, \\ S_3 &= \{p \in D \mid 1 - \beta_{\varepsilon(p)}(p) < 1 - \beta_{\varsigma(p)}(p)\} = \{p \in D \mid \beta_{\varepsilon(p)}(p) > \beta_{\varsigma(p)}(p)\}. \end{aligned}$$

and denote

$$s_1 = \#S_1, \qquad s_2 = \#S_2 \quad \text{and} \quad s_3 = \#S_3.$$

Applying Theorem 6.4, we have

$$\bar{d}_M = \left[\frac{1}{3}\Big(-n - d_1 - 2n - d_2 \right.$$
$$\left. + \sum\left(1 - \alpha(p) + 1 - \beta_{\varsigma(p)}(p) - 2 + 2\beta_{\varepsilon(p)}(p)\right) + 2g - 2\Big)\right] + 1$$
$$= -n + \left[\frac{1}{3}\Big(-\Delta - \sum\left(\alpha(p) + \beta_{\varsigma(p)}(p) - 2\beta_{\varepsilon(p)}(p)\right)\Big)\right] + 1.$$

Then

(9.1)
$$P_t(\mathcal{N}_{(2,1)}(d_1,\varpi)) = \sum_\varepsilon \operatorname*{Coeff}_{x^0}\left(\frac{(1+t)^{4g}(1+xt)^{2g}t^{2d_1-2d_2-2n+2s_2+2s_3-2\bar{d}_M}x^{\bar{d}_M}}{(1-t^2)(1-x)(1-xt^2)(1-t^{-2}x)x^{d_1-d_2-n+s_1}}\right.$$
$$\left. - \frac{(1+t)^{4g}(1+xt)^{2g}t^{2d_2+2g-2+6n-2s_3+4\bar{d}_M}x^{\bar{d}_M}}{(1-t^2)(1-x)(1-xt^2)(1-t^4x)x^{d_1-d_2-n+s_1}}\right).$$

Similarly to Lemma 8.1, we can prove

LEMMA 9.2. *The index of the critical submanifold $\mathcal{N}_{(2,1)}(d_1,\varpi)$ is*

$$\lambda_{(d_1,\varpi)} = 12g - 12 + 4n - 4d_1 + 2d_2 - 2s_0,$$

where $s_0 = \#\{\beta_i(p) \mid \beta_i(p) < \alpha(p)\} = s_1 + s_2$.

PROOF. From Proposition 2.4, $f_p = \frac{1}{2}(r^2 - \sum m_j(p)^2) = 3$, since the multiplicities are all 1. For $p \in D$, the space $P_p(L \otimes K, L \otimes K)$ consists of endomorphisms of L_p, so $\dim P_p(L \otimes K, L \otimes K) = 1$ and $P_p(E_0, E_0)$ consists of parabolic endomorphisms of $(E_0)_p$, so it has dimension 3. On the other hand, $\dim N_p(E_0, L \otimes K)$ is the dimension of the space of strongly parabolic homomorphisms from $(E_0)_p$ to $(L \otimes K)_p$, so

$$\dim N_p(E_0, L \otimes K) = \begin{cases} 0 & \text{if } 1 - \alpha(p) > 1 - \beta_1(p), \\ 1 & \text{if } 1 - \beta_1(p) > 1 - \alpha(p) > 1 - \beta_2(p), \\ 2 & \text{if } 1 - \beta_2(p) > 1 - \alpha(p). \end{cases}$$

Therefore

$$s_0 = \sum_p \dim N_p(E_0, L \otimes K) = \#\{\beta_i(p) | \beta_i(p) < \alpha(p)\}.$$

Substituting this in the formula for the Morse index of Proposition 3.11, we have

$$\begin{aligned}\lambda_{(d_1,\varpi)} &= 9(2g-2) + 6n + 2\left(5(1-g-n)+4n\right) \\ &\quad + 2\left(2(1-g)-2(d_1-2g+2)+d_2-s_0\right) \\ &= 12g-12+4n-4d_1+2d_2-2s_0.\end{aligned}$$

$\square$

Therefore the contribution of all critical submanifolds of type $(2,1)$ is given as

(9.2)
$$P_t(\Delta,(2,1)) := \sum_{d_1,\varpi} t^{\lambda(d_1,\varpi)} P_t(\mathcal{N}_{(2,1)}(d_1,\varpi))$$

$$= \sum_{d_1,\varpi}\sum_{\varepsilon} \underset{x^0}{\mathrm{Coeff}} \left(\frac{(1+t)^{4g}(1+xt)^{2g} t^{12g-12+2n-2d_1-2s_1+2s_3-2\bar{d}_M} x^{\bar{d}_M}}{(1-t^2)(1-x)(1-xt^2)(1-t^{-2}x) x^{d_1-d_2-n+s_1}} \right.$$

$$\left. - \frac{(1+t)^{4g}(1+xt)^{2g} t^{14g-14+10n-4d_1+4d_2-2s_1-2s_2-2s_3+4\bar{d}_M} x^{\bar{d}_M}}{(1-t^2)(1-x)(1-xt^2)(1-t^4x) x^{d_1-d_2-n+s_1}} \right),$$

obtained by using Lemma 9.2 and (9.1).

2. The sum for fixed (ϖ,ε).

Now we shall compute the contribution (9.2) to the Poincaré polynomial of $\mathcal{M}$ from submanifolds of type $(2,1)$. As before, we do first the sum over all possibilities of d_1 for each choice of (ϖ,ε). The condition that the moduli space $\mathcal{N}_{(2,1)}(d_1,\varpi)$ be non-empty is $2g - 2 > \mathrm{par}\mu_1 - \mathrm{par}\mu_2$. This means that

$$2g - 2 > d_1 - d_2/2 + \sum (2\alpha(p) - \beta_1(p) - \beta_2(p))/2,$$

Using that $\Delta = d_1 + d_2 + 2 - 2g$, this is translated into

$$d_2 - d_1 > 2 - 2g + \frac{\Delta}{3} + \frac{4}{3}\sum (2\alpha(p) - \beta_1(p) - \beta_2(p)).$$

But $d_2 - d_1 \equiv \Delta \pmod 2$. So $d_2 - d_1 = \bar{d}_0 + 2k$, $k \geq 0$, and

$$\bar{d}_0 = 2 - 2g + 2\left[\frac{1}{2}\left(\left[\frac{\Delta}{3} + \frac{4}{3}\sum(2\alpha(p) - \beta_1(p) - \beta_2(p))\right] - \Delta\right)\right] + \Delta.$$

In (9.2) we need to compute the terms

$$\sum t^{-2d_1} x^{d_2-d_1} = t^{-\Delta-2g+2}\sum t^{d_2-d_1} x^{d_2-d_1} = \frac{t^{\bar{d}_0-\Delta-2g+2} x^{\bar{d}_0}}{1-t^2x^2},$$

$$\sum t^{-4d_1+4d_2} x^{d_2-d_1} = \frac{t^{4\bar{d}_0} x^{\bar{d}_0}}{1-t^8x^2}.$$

Plugging this into (9.2) we get

(9.3) $P_t(\Delta,(2,1)) =$

$$= \sum_{\varpi,\varepsilon} \underset{x^0}{\mathrm{Coeff}} \left(\frac{(1+t)^{4g}(1+xt)^{2g} t^{10g-10+2n-2s_1+2s_3-2\bar{d}_M+\bar{d}_0-\Delta} x^{\bar{d}_M+\bar{d}_0+n-s_1}}{(1-t^2)(1-x)(1-xt^2)(1-t^{-2}x)(1-t^2x^2)} \right.$$

$$\left. - \frac{(1+t)^{4g}(1+xt)^{2g} t^{14g-14+10n-2s_1-2s_2-2s_3+4\bar{d}_M+4\bar{d}_0} x^{\bar{d}_M+\bar{d}_0+n-s_1}}{(1-t^2)(1-x)(1-xt^2)(1-t^4x)(1-t^8x^2)} \right).$$

3. The sum over ϖ and ε.

To perform the sum in (9.3) for all choices of (ϖ,ε), we arrange the degree and the weights to satisfy Assumption 7.6. Write $\Delta = 3k + \Delta_0$, $\Delta_0 \in \{1,2\}$. Since

$\alpha_i(p)$ are sufficiently small, we have that

$$\bar{d}_M = -n + \left[-\frac{\Delta}{3}\right] + 1 = -n - k,$$

$$\bar{d}_0 = 2 - 2g + 2\left[\frac{1}{2}\left(\left[\frac{\Delta}{3}\right] - \Delta\right)\right] + \Delta + 2 = 2 - 2g + k + \Delta_0,$$

are independent of (ϖ, ε). Therefore to do the sum in (9.3), we only need to compute

$$\sum_{\varpi,\varepsilon} t^{-2s_1+2s_3} x^{-s_1} \quad \text{and} \quad \sum_{\varpi,\varepsilon} t^{-2s_1-2s_2-2s_3} x^{-s_1}.$$

As before,

$$s_1 = \sum_{p \in D} s_1(p),$$

$$s_2 = \sum_{p \in D} s_2(p),$$

$$s_3 = \sum_{p \in D} s_3(p),$$

where $s_1(p)$, $s_2(p)$ and $s_3(p)$ are defined in the obvious way:

$$s_1(p) = \begin{cases} 1 & \text{if } \alpha(p) > \beta_{\varsigma(p)}(p), \\ 0 & \text{otherwise,} \end{cases}$$

$$s_2(p) = \begin{cases} 1 & \text{if } \alpha(p) > \beta_{\varepsilon(p)}(p), \\ 0 & \text{otherwise,} \end{cases}$$

and

$$s_3(p) = \begin{cases} 1 & \text{if } \beta_{\varepsilon(p)}(p) > \beta_{\varsigma(p)}(p), \\ 0 & \text{otherwise.} \end{cases}$$

The values of $s_1(p)$, $s_2(p)$ and $s_3(p)$ as a function of $(\varpi(p), \varepsilon(p))$ are in Table 9.1,

TABLE 9.1. $s_1(p)$, $s_2(p)$ and $s_3(p)$ as a function of $(\varpi(p), \varepsilon(p))$

$(\varpi(p), \varepsilon(p))$	$(1,1)$	$(1,2)$	$(2,1)$	$(2,2)$	$(3,1)$	$(3,2)$
$s_1(p)$	0	0	0	1	1	1
$s_2(p)$	0	0	1	0	1	1
$s_3(p)$	0	1	0	1	0	1

We obtain

$$\sum_{\varpi,\varepsilon} t^{-2s_1+2s_3} x^{-s_1} = \prod_{p \in D} \sum_{\varpi(p),\varepsilon(p)} t^{-2s_1(p)+2s_3(p)} x^{-s_1(p)}$$

$$= (t^{-2}x^{-1} + 2 + 2x^{-1} + t^2)^n$$

$$= t^{-2n} x^{-n} (1 + 2t^2 + 2t^2 x + t^4 x)^n,$$

and
$$\sum_{\varpi,\varepsilon} t^{-2s_1-2s_2-2s_3} x^{-s_1} = (2t^{-4}x^{-1} + t^{-6}x^{-1} + 1 + 2t^{-2})^n$$
$$= t^{-6n} x^{-n} (1 + 2t^2 + 2t^4 x + t^6 x)^n.$$

Plugging this into (9.3) we get that $P_t(\Delta, (2,1))$ equals

$$\operatorname*{Coeff}_{x^0} \left(\frac{(1+t)^{4g}(1+xt)^{2g} t^{10g-10-2\bar{d}_M + \bar{d}_0 - \Delta} x^{\bar{d}_M + \bar{d}_0} (1 + 2t^2 + 2t^2 x + t^4 x)^n}{(1-t^2)(1-x)(1-xt^2)(1-t^{-2}x)(1-t^2 x^2)} \right.$$
$$\left. - \frac{(1+t)^{4g}(1+xt)^{2g} t^{14g-14+4n+4\bar{d}_M + 4\bar{d}_0} x^{\bar{d}_M + \bar{d}_0} (1 + 2t^2 + 2t^4 x + t^6 x)^n}{(1-t^2)(1-x)(1-xt^2)(1-t^4 x)(1-t^8 x^2)} \right).$$

Now, using that $\bar{d}_M + \bar{d}_0 = 2 - 2g - n + \Delta_0$ and $-2\bar{d}_M + \bar{d}_0 - \Delta = 2 - 2g + 2n$, we obtain the following.

PROPOSITION 9.3. *Under Assumption 7.6, let $\Delta_0 \in \{1,2\}$ be the remainder modulo 3 of Δ. Then $P_t(\Delta,(2,1))$ equals*

$$\operatorname*{Coeff}_{x^0} \left(\frac{(1+t)^{4g}(1+xt)^{2g} t^{8g-8+2n} x^{2-2g+\Delta_0-n} (1 + 2t^2 + 2t^2 x + t^4 x)^n}{(1-t^2)(1-x)(1-xt^2)(1-t^{-2}x)(1-t^2 x^2)} \right.$$
$$\left. - \frac{(1+t)^{4g}(1+xt)^{2g} t^{6g-6+4\Delta_0} x^{2-2g+\Delta_0-n} (1 + 2t^2 + 2t^4 x + t^6 x)^n}{(1-t^2)(1-x)(1-xt^2)(1-t^4 x)(1-t^8 x^2)} \right).$$

□

CHAPTER 10

Betti numbers of the moduli space of rank three parabolic bundles

The Betti numbers of the moduli space of parabolic vector bundles were computed by Nitsure [34] and Holla [25]. Here we work out Holla's formula for the special case when the rank is 3 and all flags at the parabolic points are full. We also continue to work with the choice of weights made in Assumptions 5.1 and 7.6.

1. Notation

Given a parabolic bundle E, the corresponding *quasi-parabolic data*, R, gives the multiplicity of each step of the flag at the parabolic points:

$$R_i^p = \dim E_{p,i} - \dim E_{p,i+1}$$

where $E_p = E_{p,1} \supset \cdots \supset E_{p,s_p+1} = 0$ is the parabolic filtration at p. Thus $R_i^p = m_i(p)$ in the notation of Section 1 of Chapter 2. We choose to keep Holla's original notation in this chapter because it is better suited for the calculations to be carried out. The *rank* of R is just the rank of E, $n(R) = \sum_i R_i^p$. One defines $\alpha(R) = \sum_{p,i} \alpha_i(p) R_i^p$, so that

$$\mathrm{pardeg}(E) = \deg(E) + \alpha(R).$$

Given a parabolic bundle E with Harder–Narasimhan filtration

$$0 = G_0 \subsetneq G_1 \subsetneq \cdots \subsetneq G_r = E,$$

each subbundle G_j is a parabolic bundle with the induced parabolic structure. The induced quasi-parabolic data $R^I_{\leq k}$ is defined by

$$(R^I_{\leq k})_i^p = \dim(G_{k,p} \cap E_{p,i}) - \dim(G_{k,p} \cap E_{p,i+1}).$$

Thus $(R^I_{\leq k})_i^p$ is the multiplicity of the i-th step of the induced parabolic structure on G_k at p (note that this may be zero). Each subquotient G_k/G_{k-1} is also a parabolic bundle and the corresponding parabolic data is R^I_k, given by

$$(R^I_k)_i^p = (R^I_{\leq k})_i^p - (R^I_{\leq k-1})_i^p.$$

The *intersection matrix* I is defined by letting

$$I_{i,k}^p = (R^I_k)_i^p,$$

in other words, $I_{i,k}^p$ is the multiplicity of the i-th step of the induced parabolic structure on G_k/G_{k-1} at p. The rank of the subquotient G_k/G_{k-1} is $n(R^I_k) = \sum_i I_{i,k}^p$ and hence the Harder–Narasimhan type of E can be written as

$$\mathbf{n} = (n_1, \ldots, n_r) = \big(n(R^I_1), \ldots, n(R^I_r)\big).$$

2. Holla's formula

The formula [**25**, Theorem 5.23] for the Poincaré polynomial of the moduli space of parabolic bundles of degree Δ and rank $n(R)$ is

(10.1)

$$P_t(\Delta, n(R)) =$$

$$= (1-t^2) \sum_{r=1}^{n(R)} \sum_{I} \frac{t^{2\{\sigma'(I) - \Delta(n(R) - n(R_r^I)) + M_g(I,\alpha)\}}}{(t^{2n(R_1^I) + 2n(R_2^I)} - 1) \cdots (t^{2n(R_{r-1}^I) + 2n(R_r^I)} - 1)} \prod_{k=1}^{r} P_{R_k^I}(t),$$

where the sum is over intersection matrices I of all possible Harder–Narasimhan filtrations of parabolic bundles. Here r is the length of the Harder–Narasimhan filtration corresponding to I, the number $\sigma'(I)$ is defined by $\sigma'(I) = \sum_{p \in D} \sigma'_p(I)$, where

$$\sigma'_p(I) = \sum_{k > l,\ i < j} I^p_{i,k} I^p_{j,l},$$

the number $M_g(I, \alpha)$ is defined by

$$M_g(I, \alpha) = \sum_{k=1}^{r-1} \left(n(R_k^I) + n(R_{k+1}^I) \right) \left(\left[n(R_{\leq k}^I) \frac{\Delta + \alpha(R)}{n(R)} - \alpha(R_{\leq k}^I) \right] + 1 \right)$$

$$+ (g-1) \sum_{i<j} n(R_i^I) n(R_j^I)$$

and

$$P_R(t) = \left(\frac{\prod_{i=1}^{n(R)} (1 - t^{2i})^n}{\prod_{p \in D} \prod_{\{i \mid R_i^p \neq 0\}} \prod_{l=1}^{R_i^p} (1 - t^{2l})} \right) \left(\frac{\prod_{i=1}^{n(R)} (1 + t^{2i-1})^{2g}}{(1 - t^{2n(R)}) \prod_{i=1}^{n(R)-1} (1 - t^{2i})^2} \right).$$

This formula is valid for all choices of weights such that a parabolically semistable bundle is automatically parabolically stable. In particular, it is valid under our Assumption 5.1 on genericity of the weights.

REMARK 10.1. This is the formula for the non-fixed determinant case, whereas Holla states the formula for the fixed determinant case. The two formulas differ by a factor of $(1+t)^{2g}$ coming from the Poincaré polynomial of the Jacobian (see [**34**]).

3. The rank 3 case

We now work out explicitly Holla's formula for the case of rank 3 parabolic bundles under Assumptions 5.1 and 7.6. This implies in particular that all parabolic flags are full. For full flags we have the following simplification of the expressions $P_{R_k^I}(t)$.

PROPOSITION 10.2. Assume that all parabolic flags are full. Then

$$P_{R_k^I}(t) = \frac{\prod_{i=1}^{n(R_k^I)} (1 - t^{2i})^{n-1} (1 + t^{2i-1})^{2g}}{(1 - t^2)^{n(R_k^I)n} \prod_{i=1}^{n(R_k^I)-1} (1 - t^{2i})}.$$

PROOF. Since all flags are full, we have $\#\{i \mid (R_k^I)_i^p \neq 0\} = n(R_k^I)$. □

3. THE RANK 3 CASE

Note that $P_{R_k^I}(t)$ only depends on I through the rank $n(R_k^I)$ of G_k/G_{k-1}. For $n_k = n(R_k^I)$ we shall therefore write

$$P_{n_k}(t) = P_{R_k^I}(t).$$

We calculate P_{n_k} for $n_k = 1, 2, 3$. We obtain

(10.2)
$$P_1(t) = \frac{(1+t)^{2g}}{(1-t^2)},$$
$$P_2(t) = \frac{(1+t^2)^{n-1}(1+t)^{2g}(1+t^3)^{2g}}{(1-t^2)^3},$$
$$P_3(t) = \frac{(1+2t^2+2t^4+t^6)^{n-1}(1+t)^{2g}(1+t^3)^{2g}(1+t^5)^{2g}}{(1-t^2)^4(1-t^4)}.$$

Now rewrite (10.1) as

(10.3)
$$P_t(\Delta, n(R)) = (1-t^2) \sum_{\mathbf{n}} \sum_{I \text{ of type } \mathbf{n}} t^{2\sigma'(I)} Q_I(t) \prod_{k=1}^r P_{n_k}(t),$$

where

$$Q_I(t) = \frac{t^{2\{M_g(I,\alpha)-\Delta(n(R)-n(R_r^I))\}}}{(t^{2n(R_1^I)+2n(R_2^I)}-1)\cdots(t^{2n(R_{r-1}^I)+2n(R_r^I)}-1)}.$$

For rank $n(R) = 3$, the possible Harder–Narasimhan types $\mathbf{n}$ are (3), $(1,2)$, $(2,1)$ and $(1,1,1)$. In the following, we list all the possible intersection matrices I according to the various types for rank 3. We also give the corresponding values of $\sigma_p'(I)$.

Intersection matrix for type (3).

$I_{i,k}^p$	$k=1$
$i=1$	1
$i=2$	1
$i=3$	1
$\sigma_p'(I)$	0

Intersection matrices for type $(1,2)$.

$I_{i,k}^p$	$k=1$	$k=2$
$i=1$	1	0
$i=2$	0	1
$i=3$	0	1
$\sigma_p'(I)$	0	

$I_{i,k}^p$	$k=1$	$k=2$
$i=1$	0	1
$i=2$	1	0
$i=3$	0	1
$\sigma_p'(I)$	1	

$I_{i,k}^p$	$k=1$	$k=2$
$i=1$	0	1
$i=2$	0	1
$i=3$	1	0
$\sigma_p'(I)$	2	

Intersection matrices for type $(2,1)$.

$I_{i,k}^p$	$k=1$	$k=2$
$i=1$	1	0
$i=2$	1	0
$i=3$	0	1
$\sigma_p'(I)$	0	

$I_{i,k}^p$	$k=1$	$k=2$
$i=1$	1	0
$i=2$	0	1
$i=3$	1	0
$\sigma_p'(I)$	1	

$I_{i,k}^p$	$k=1$	$k=2$
$i=1$	0	1
$i=2$	1	0
$i=3$	1	0
$\sigma_p'(I)$	2	

Intersection matrices for type $(1,1,1)$.

$I^p_{i,k}$	$k=1$	$k=2$	$k=3$
$i=1$	1	0	0
$i=2$	0	1	0
$i=3$	0	0	1
$\sigma'_p(I)$		0	

$I^p_{i,k}$	$k=1$	$k=2$	$k=3$
$i=1$	1	0	0
$i=2$	0	0	1
$i=3$	0	1	0
$\sigma'_p(I)$		1	

$I^p_{i,k}$	$k=1$	$k=2$	$k=3$
$i=1$	0	1	0
$i=2$	1	0	0
$i=3$	0	0	1
$\sigma'_p(I)$		1	

$I^p_{i,k}$	$k=1$	$k=2$	$k=3$
$i=1$	0	1	0
$i=2$	0	0	1
$i=3$	1	0	0
$\sigma'_p(I)$		2	

$I^p_{i,k}$	$k=1$	$k=2$	$k=3$
$i=1$	0	0	1
$i=2$	1	0	0
$i=3$	0	1	0
$\sigma'_p(I)$		2	

$I^p_{i,k}$	$k=1$	$k=2$	$k=3$
$i=1$	0	0	1
$i=2$	0	1	0
$i=3$	1	0	0
$\sigma'_p(I)$		3	

Now we compute the exponent of the power of t^2 in the numerator of $Q_I(t)$. This is greatly simplified thanks to our assumption on the degree and weights.

PROPOSITION 10.3. *Let* $n(R) = 3$. *Under Assumption 7.6, we have*

$$M_g(I,\alpha) - \Delta(n(R) - n(R_r^I)) = \begin{cases} 3\left[\frac{\Delta}{3}\right] - \Delta + 2g + 1, & \text{for } I \text{ of type } (1,2), \\ 3\left[\frac{2\Delta}{3}\right] - 2\Delta + 2g + 1, & \text{for } I \text{ of type } (2,1), \\ 3g - 1, & \text{for } I \text{ of type } (1,1,1), \end{cases}$$

PROOF. As $\Delta \not\equiv 0 \pmod{3}$, we have that $n(R^I_{\leq k})\frac{\Delta}{n(R)}$ is non-integer. Since the weights are small, we have that, for $k \leq r-1$,

$$\left[n(R^I_{\leq k})\frac{\Delta + \alpha(R)}{n(R)} - \alpha(R^I_{\leq k})\right] + 1 = \left[n(R^I_{\leq k})\frac{\Delta}{n(R)}\right] + 1.$$

Substituting into the definition of $M_g(I,\alpha)$, we get the result. For type $(1,1,1)$, we have used that $\left[\frac{\Delta}{3}\right] + \left[\frac{2\Delta}{3}\right] - \Delta = -1$, since $\Delta \not\equiv 0 \pmod{3}$. □

Note that, in particular, Proposition 10.3 implies that $Q_I(t)$ only depends on I through its Harder–Narasimhan type. We shall therefore need to calculate $\sum_I t^{2\sigma'(I)}$ for each type. This is an easy task using the tables given for the intersection matrices and the fact that $\sum_I t^{2\sigma'(I)} = \prod_{p \in D} \sum_{I^p} t^{2\sigma'_p(I)}$, as is easily seen by induction on the number of points in D. The result is:

(10.4)
$$\sum_I t^{2\sigma'(I)} = (1 + t^2 + t^4)^n, \qquad \text{for } I \text{ of type } (1,2) \text{ and } (2,1),$$
$$\sum_I t^{2\sigma'(I)} = (1 + 2t^2 + 2t^4 + t^6)^n, \quad \text{for } I \text{ of type } (1,1,1).$$

We can now calculate the contribution to $P_t(\Delta, 3)$ from I of type $(1,1,1)$:

(10.5)
$$\sum_{I \text{ of type } (1,1,1)} t^{2\sigma'(I)} Q_I(t) \prod_{k=1}^r P_{n_k}(t) = \frac{(1 + 2t^2 + 2t^4 + t^6)^n t^{6g-2}}{(t^4 - 1)^2} P_1(t)^3.$$

Similarly, the contributions to $P_t(\Delta,3)$ from I of type $(1,2)$ and $(2,1)$ are:

$$\sum_{I \text{ of type } (1,2)} t^{2\sigma'(I)}Q_I(t)\prod_{k=1}^{r}P_{n_k}(t) = \frac{(1+t^2+t^4)^n t^{2\{3[\frac{\Delta}{3}]-\Delta+2g+1\}}}{t^6-1}P_1(t)P_2(t),$$

$$\sum_{I \text{ of type } (2,1)} t^{2\sigma'(I)}Q_I(t)\prod_{k=1}^{r}P_{n_k}(t) = \frac{(1+t^2+t^4)^n t^{2\{3[\frac{2\Delta}{3}]-2\Delta+2g+1\}}}{t^6-1}P_1(t)P_2(t).$$

Summing the contributions of type $(1,2)$ and type $(2,1)$ some simplification results because, whenever $\Delta \not\equiv 0 \pmod 3$, one has

$$t^{2\{3[\frac{\Delta}{3}]-\Delta+2g+1\}} + t^{2\{3[\frac{2\Delta}{3}]-2\Delta+2g+1\}} = t^{4g+2}(t^{2\{3[\frac{\Delta}{3}]-\Delta\}} + t^{2\{3[\frac{2\Delta}{3}]-2\Delta\}})$$
$$= t^{4g+2}(t^{-2}+t^{-4})$$
$$= t^{4g-2}(1+t^2).$$

Hence we obtain
(10.6)
$$\sum_{I \text{ of type } (1,2) \text{ or } (2,1)} t^{2\sigma'(I)}Q_I(t)\prod_{k=1}^{r}P_{n_k}(t) = \frac{(1+t^2+t^4)^n t^{4g-2}(1+t^2)}{t^6-1}P_1(t)P_2(t).$$

Since for $\mathbf{n}=(3)$ we clearly have
(10.7)
$$t^{2\sigma'(I)}Q_I(t) = 1,$$

we are now in a position to put everything together and calculate $P_t(\Delta,3)$ for $\Delta \not\equiv 0 \pmod 3$.

PROPOSITION 10.4. *Under Assumption 7.6, the Poincaré polynomial of the moduli space of stable parabolic bundles of rank 3 is given by*

$$P_t(\Delta,3) = \frac{(1+t)^{2g}(1+2t^2+2t^4+t^6)^{n-1}}{(1-t^2)^3(1-t^4)} \cdot$$
$$\cdot \Big((1+t^3)^{2g}(1+t^5)^{2g} + (1+t)^{4g}(1+t^2+t^4)t^{6g-2}$$
$$- (1+t)^{2g}(1+t^3)^{2g}(1+t^2)^2 t^{4g-2}\Big).$$

PROOF. Substituting (10.5), (10.6) and (10.7) in (10.3) and using (10.2) we obtain

$$P_t(\Delta,3) = (1-t^2)\Big(P_3(t) + \frac{(1+t^2+t^4)^n t^{4g-2}(1+t^2)}{t^6-1}P_1(t)P_2(t)$$
$$+ \frac{(1+2t^2+2t^4+t^6)^n t^{6g-2}}{(t^4-1)^2}P_1(t)^3\Big)$$
$$= \frac{(1+2t^2+2t^4+t^6)^{n-1}(1+t)^{2g}(1+t^3)^{2g}(1+t^5)^{2g}}{(1-t^2)^3(1-t^4)}$$
$$+ \frac{(1+t^2+t^4)^n t^{4g-2}(1+t^2)(1+t)^{2g}(1+t^2)^{n-1}(1+t)^{2g}(1+t^3)^{2g}}{(t^6-1)(1-t^2)^3}$$
$$+ \frac{(1+2t^2+2t^4+t^6)^n t^{6g-2}(1+t)^{6g}}{(1-t^4)^2(1-t^2)^2}.$$

Simplifying this expression we obtain the formula stated. □

CHAPTER 11

Betti numbers of the moduli space of rank three parabolic Higgs bundles

In this chapter we put everything together to obtain the Poincaré polynomial of the moduli space of rank three parabolic Higgs bundles.

1. Poincaré polynomial

THEOREM 11.1. *Let $\mathcal{M}$ be the moduli space of rank three parabolic Higgs bundles of some fixed degree and weights, over a connected, smooth projective complex algebraic curve of genus g. If the weights are generic (in the sense that there are no properly semistable parabolic Higgs bundles), then the Poincaré polynomial of $\mathcal{M}$ is given by*

$$P_t(\mathcal{M}) = \underset{u^0 v^0}{\mathrm{Coeff}} \left((1 + 2u^2vt^2 + 2uv^2t^2 + u^3v^3t^4)^n \cdot \right.$$

$$\left. \cdot \frac{t^{2(4g-3+n)}(1+t)^{2g}(1+u^2vt)^{2g}(1+uv^2t)^{2g}}{u^{3n+6g-8}v^{3n+6g-7}(1-u^2v)(1-uv^2)(1-u^2vt^2)(1-uv^2t^2)(1-v^3t^2)(1-u^3t^2)} \right)$$

$$+ \underset{x^0}{\mathrm{Coeff}} \left(\frac{(1+t)^{4g}(1+xt)^{2g}t^{6g-6}x^{2-2g-n}}{(1-t^2)(1-x)(1-xt^2)} \cdot \right.$$

$$\left. \cdot \left(\frac{t^{2g-2+2n}(x+x^2)(1+2t^2+2t^2x+t^4x)^n}{(1-t^{-2}x)(1-t^2x^2)} - \frac{(t^4x+t^8x^2)(1+2t^2+2t^4x+t^6x)^n}{(1-t^4x)(1-t^8x^2)} \right) \right)$$

$$+ \frac{(1+t)^{2g}(1+2t^2+2t^4+t^6)^{n-1}}{(1-t^2)^3(1-t^4)} \cdot$$

$$\cdot \left((1+t^3)^{2g}(1+t^5)^{2g} + (1+t)^{4g}(1+t^2+t^4)t^{6g-2} \right.$$

$$\left. - (1+t)^{2g}(1+t^3)^{2g}(1+t^2)^2 t^{4g-2} \right).$$

PROOF. It follows from Morse theory, as explained in Chapter 3, that $P_t(\mathcal{M}) = P_t(\Delta, 3) + P_t(\Delta, (1, 2)) + P_t(\Delta, (2, 1)) + P_t(\Delta, (1, 1, 1))$, where the polynomials on the right hand side are given in Propositions 7.7, 8.2, 9.3 and 10.4. In order to apply these formulas we need to choose $\Delta_0 \in \{1, 2\}$. However, the contribution $P_t(\Delta, (1, 1, 1))$ of Proposition 7.7 is independent of this choice, by using the duality $(u, v) \mapsto (v, u)$. Also, the contribution $P_t(\Delta, (1, 2)) + P_t(\Delta, (2, 1))$ from Propositions 8.2 and 9.3 is independent of the choice of Δ_0 by using that for $\Delta_0 = 1, 2$ we have $x^{\Delta_0} + x^{3-\Delta_0} = x + x^2$ and $x^{\Delta_0}t^{4\Delta_0} + x^{3-\Delta_0}t^{12-4\Delta_0} = t^4x + t^8x^2$.

Even though the various contributions to the Poincaré polynomial were calculated for a specific choice of degree and weights (cf. Assumptions 5.1 and 7.6), we know by Proposition 2.1 that the final result is independent of this choice. □

REMARK 11.2. Obviously, it is possible to do the computation of $P_t(\mathcal{M})$ under different choices of degree and weights. Most of the calculations in Chapters 7–11 are carried out in general, and we have always introduced our Assumption 7.6 as late as possible in each chapter. Of course, the final answer will be the same as the one given in Theorem 11.1, though the partial contributions of the critical submanifolds of different types may differ.

2. Special low genus cases

We can calculate the Poincaré polynomial of the moduli space of parabolic Higgs bundles for specific values of n and g by using a computer algebra system. For instance, if $n = 1$ and $g = 2$ then Theorem 11.1 gives

$$\begin{aligned}P_t(\mathcal{M}) = {}& 36\,t^{26} + 324\,t^{25} + 1368\,t^{24} + 3620\,t^{23} + 6810\,t^{22} + 9860\,t^{21} + 11670\,t^{20} \\ & + 11876\,t^{19} + 10860\,t^{18} + 9224\,t^{17} + 7408\,t^{16} + 5688\,t^{15} + 4216\,t^{14} \\ & + 3036\,t^{13} + 2134\,t^{12} + 1464\,t^{11} + 981\,t^{10} + 640\,t^9 + 401\,t^8 + 244\,t^7 \\ & + 144\,t^6 + 80\,t^5 + 42\,t^4 + 20\,t^3 + 9\,t^2 + 4\,t + 1\,,\end{aligned}$$

and when $n = 2$ and $g = 2$ we obtain

$$\begin{aligned}P_t(\mathcal{M}) = {}& 252\,t^{32} + 2416\,t^{31} + 10848\,t^{30} + 30540\,t^{29} + 61178\,t^{28} + 94368\,t^{27} \\ & + 119187\,t^{26} + 129952\,t^{25} + 127737\,t^{24} + 116656\,t^{23} + 100849\,t^{22} \\ & + 83564\,t^{21} + 66925\,t^{20} + 52100\,t^{19} + 39605\,t^{18} + 29504\,t^{17} + 21572\,t^{16} \\ & + 15472\,t^{15} + 10884\,t^{14} + 7496\,t^{13} + 5043\,t^{12} + 3312\,t^{11} + 2113\,t^{10} \\ & + 1308\,t^9 + 782\,t^8 + 448\,t^7 + 247\,t^6 + 128\,t^5 + 62\,t^4 + 28\,t^3 + 11\,t^2 \\ & + 4\,t + 1\,.\end{aligned}$$

Another example is the Poincaré polynomial of $\mathcal{M}$ for $g = 1$ and $n = 1$,

$$P_t(\mathcal{M}) = 6\,t^8 + 18\,t^7 + 24\,t^6 + 20\,t^5 + 13\,t^4 + 8\,t^3 + 4\,t^2 + 2\,t + 1\,.$$

REMARK 11.3. In [20] Hausel conjectured a formula for the Poincaré polynomial of the moduli space of stable Higgs bundles of any rank. Hausel has informed us of an analogous conjecture for the Poincaré polynomial of the moduli space of *parabolic* Higgs bundles of any rank: in the case of rank three the mixed Hodge polynomial of the corresponding character variety is

$$\begin{aligned}H_3^n(q,t) = {}& \frac{((qt^2+1)(q^2t^4+qt^2+1))^n (q^3t^5+1)^{2g}(q^2t^3+1)^{2g}}{(q^3t^6-1)(q^3t^4-1)(q^2t^4-1)(q^2t^2-1)} \\ & + \frac{(q^3t^6(q+1)(q^2+q+1))^n q^{6g-6}t^{12g-12}(q^3t+1)^{2g}(q^2t+1)^{2g}}{(q^3t^2-1)(q^3-1)(q^2t^2-1)(q^2-1)} \\ & + \frac{(q^2t^4(2q^2t^2+qt^2+q+2))^n q^{4g-4}t^{8g-8}(q^3t^3+1)^{2g}(qt+1)^{2g}}{(q^3t^4-1)(q^3t^2-1)(qt^2-1)(q-1)} \\ & + \frac{6^n (qt^2)^{3n} q^{6g-6} t^{12g-12}(qt+1)^{4g}}{3(qt^2-1)^2(q-1)^2} \\ & - \frac{(3q^2t^4(qt^2+1))^n q^{4g-4}t^{8g-8}(q^2t^3+1)^{2g}(qt+1)^{2g}}{(q^2t^4-1)(q^2t^2-1)(qt^2-1)(q-1)} \\ & - \frac{(3q^3t^6(q+1))^n q^{6g-6}t^{12g-12}(q^2t+1)^{2g}(qt+1)^{2g}}{(q^2t^2-1)(q^2-1)(qt^2-1)(q-1)}\,.\end{aligned}$$

Hausel conjectures that the Poincaré polynomial of the moduli space of rank three parabolic Higgs bundles with n marked points and full flags is obtained from this polynomial by the substitution $P_t(\mathcal{M}) = H_3^n(1,t)$. The formulas are difficult to compare in general, but in computer calculations Hausel's formula provides the same result as ours in all the cases that we have checked. Thus our results provide evidence for this conjecture.

We finish by considering the case when X has genus zero. It is easy to see that under our Assumption 7.6 of small weights, the moduli space of stable parabolic bundles on X is empty, because any parabolically stable bundle would have to be stable. However, our results show that there are non-empty critical submanifolds of the moduli space of parabolic Higgs bundles for $n \geq 3$: for example, if $g = 0$ and $n = 3$, our calculations show that

$$P_t(\mathcal{M}) = 7t^2 + 1,$$

where the only contribution is from critical submanifolds of type $(1,1,1)$. This means that stable parabolic Higgs bundles exist and hence the moduli space is non-empty. Following our general description of the critical submanifolds of Section 1 of Chapter 7 one can explicitly describe the critical submanifolds of type $(1,1,1)$ and thus get examples of stable parabolic Higgs bundles. One such example is given as follows. Define parabolic line bundles $L_1 = \mathcal{O}(1)$, $L_2 = \mathcal{O}$ and $L_3 = \mathcal{O}$ with small weights $\alpha_i(p)$ on L_i, such that $\alpha_1(p) < \alpha_2(p) < \alpha_3(p)$ at each marked point. Let $E = L_1 \oplus L_2 \oplus L_3$. Then any map from L_i to L_{i+1} is strongly parabolic and we can define a Higgs field Φ with non-zero components in $\mathrm{SParHom}(L_1, L_2 \otimes K(D)) = \mathrm{Hom}(L_1, L_2 \otimes K(D)) = \Gamma(\mathcal{O}(n-3))$ and $\mathrm{SParHom}(L_2, L_3 \otimes K(D)) = \mathrm{Hom}(L_2, L_3 \otimes K(D)) = \Gamma(\mathcal{O}(n-2))$. Clearly the resulting parabolic Higgs bundle is stable. When $n = 3$ this parabolic Higgs bundle is a minimum of the Morse function, as follows from Lemma 7.4, and the critical submanifold consisting of such parabolic Higgs bundles is easily seen to be isomorphic to $\mathbb{P}^1$. From our description of the critical submanifolds of Section 1 of Chapter 7 one sees that there are six other critical submanifolds, all consisting of parabolic Higgs bundles with the same underlying vector bundle but with different distributions of the weights. All these other critical submanifolds consist of one point and have index 2. Of course these observations check with our calculation of the Poincaré polynomial.

One can give a very explicit description of the moduli space in the $n = 3$ and $\Delta = 0$ case (this is of course different from the $\Delta = 1$ moduli space considered in the previous paragraph but, as we know, has the same Betti numbers). This is done by means of the Hitchin map ([**24**]), which exhibits the moduli space as an elliptic fibration over $\mathbb{C}$ (in fact an ALG manifold [**10**]). To carry this out, consider the general case where the bundle over $\mathbb{P}^1$ is trivial and the three points are 0, 1 and ∞. The Higgs field (twisting by $K(3) = \mathcal{O}(1)$) can be written as

$$\Phi = Az + B(z-1)$$

where A, B and $A + B$ are nilpotent since these are the residues at the parabolic points. That means that $\mathrm{Tr}\,\Phi = 0$, since $\mathrm{Tr}\,A = \mathrm{Tr}\,B = 0$, and $\mathrm{Tr}\,\Phi^2 = 0$ since $\mathrm{Tr}\,A^2 = \mathrm{Tr}\,B^2 = 0$ and $\mathrm{Tr}\,(A+B)^2 = 0$, which means $\mathrm{Tr}\,AB = 0$; also $\mathrm{Tr}\,\Phi^3 = cz(z-1)$ using that $\mathrm{Tr}\,A^3 = \mathrm{Tr}\,B^3 = 0$ and $\mathrm{Tr}\,(A+B)^3 = 0$. The spectral curve ([**24**]) has the form

$$w^3 = kz(z-1)$$

which is a cubic curve invariant by $\mathbb{Z}/3$ by multiplying w by a cube root of unity. As k varies in $\mathbb{C}$ we have the elliptic fibration with an E_6 curve at $k=0$. Of course the Hitchin map is just $(E, \Phi) \mapsto k$ and, in particular, the nilpotent cone is the E_6 curve.

For higher values of n, there are also contributions from critical submanifolds of type $(1,2)$ and $(2,1)$. For instance, for $g=0$ and $n=4$ our formula gives

$$P_t(\mathcal{M}) = 271\, t^8 + 144\, t^6 + 43\, t^4 + 9\, t^2 + 1\,,$$

with non-zero contributions from critical submanifolds of type $(1,2)$. For $g=0$ and $n=5$ critical submanifolds of both type $(1,2)$ and $(2,1)$ contribute and one obtains

$$P_t(\mathcal{M}) = 4645\, t^{14} + 3791\, t^{12} + 1926\, t^{10} + 762\, t^8 + 249\, t^6 + 63\, t^4 + 11\, t^2 + 1\,.$$

CHAPTER 12

The fixed determinant case

The goal of this chapter is to calculate the Poincaré polynonial of the moduli space of rank 3 parabolic Higgs bundles with fixed determinant. We follow our calculation for the non-fixed determinant case closely and only point out the main differences. The final result is given in Theorem 12.20. As a corollary we obtain the fact that fixed determinant moduli space has Euler characteristic zero—note that this is not the case for the usual fixed determinant Higgs bundle moduli space, cf. [23], [16] and [20].

1. Preliminaries

Let E be a rank r parabolic bundle with degree Δ and weights $\alpha_i(p)$ with multiplicities $m_i(p)$. Then the determinant $\Lambda^r E$ is a parabolic bundle, with degree $\bar{\Delta} = \Delta + \sum_{p \in D} \left[\sum_i m_i(p)\alpha_i(p) \right]$ and weights $\sum_i m_i(p)\alpha_i(p) - \left[\sum_i m_i(p)\alpha_i(p) \right]$, for $p \in D$ (in particular, under Assumption 7.6, we have $\bar{\Delta} = \Delta$). Now, for any choice of weights, the moduli space of rank 1 parabolic Higgs bundles of degree $\bar{\Delta}$ is naturally identified with the total space of the cotangent bundle to the Jacobian of degree $\bar{\Delta}$ line bundles on X. Consider the "determinant map" from the moduli space of stable rank r parabolic Higgs bundles $\mathcal{M}$ to $T^* \mathrm{Jac}^{\bar{\Delta}}(X)$:

(12.1)
$$\det \colon \mathcal{M} \to T^* \mathrm{Jac}^{\bar{\Delta}}(X),$$
$$(E, \Phi) \mapsto (\Lambda^r E, \mathrm{Tr}\, \Phi).$$

Note that $T^* \mathrm{Jac}^l(X) \cong \mathrm{Jac}^l(X) \times H^1(\mathcal{O})$ in a canonical way.

Let Λ be a fixed line bundle of degree $\bar{\Delta}$. By definition, the fibre of det over $(\Lambda, 0)$ is the moduli space of stable parabolic Higgs bundles with fixed determinant Λ:

$$\mathcal{M}^\Lambda = \det^{-1}(\Lambda, 0).$$

We shall need the following analogue of Proposition 2.1: it is not hard to see that the proof, including the relevant parts of [38], goes over to the fixed determinant case.

PROPOSITION 12.1. *Fix the rank r. For different choices of the determinant bundle Λ and generic weights, the moduli spaces $\mathcal{M}^\Lambda$ have the same Betti numbers.*
□

REMARK 12.2. The group of r-torsion points in the Jacobian, $\Gamma_r = \{L \mid L^r = \mathcal{O}\}$, acts on $\mathcal{M}^\Lambda$ by tensor product:

$$(E, \Phi) \mapsto (E \otimes L, \Phi).$$

We also have an action of Γ_r on $T^* \mathrm{Jac}^l(X)$ given by

$$(M, \alpha) \mapsto (M \otimes L^{-1}, \alpha),$$

for $L \in \Gamma_r$ and, via this action, the covering
$$T^* \operatorname{Jac}^l(X) \to T^* \operatorname{Jac}^{r\,l}(X)\,,$$
$$(M, \alpha) \mapsto (M^r, \alpha)\,,$$
can be viewed as a principal Γ_r-bundle. As done in Atiyah–Bott [**2**] for ordinary bundles, we can use this to express $\mathcal{M}$ as a fibred product
$$\mathcal{M}^\Lambda \times_{\Gamma_r} T^* \operatorname{Jac}^0(X) \xrightarrow{\cong} \mathcal{M}\,,$$
$$\bigl((E, \Phi), (L, \alpha)\bigr) \mapsto (E \otimes L, \Phi + \alpha \operatorname{Id})\,.$$
It follows that the rational cohomology of $\mathcal{M}$ is isomorphic to the Γ_r-invariant part of the cohomology of $\mathcal{M}^\Lambda \times_{\Gamma_r} T^* \operatorname{Jac}^0(X)$. But Γ_r acts trivially on the cohomology of $T^* \operatorname{Jac}^0(X)$ and, therefore,
$$H^*(\mathcal{M}; \mathbb{Q}) \cong H^*(\mathcal{M}^\Lambda; \mathbb{Q})^{\Gamma_r} \otimes H^*(\operatorname{Jac}^0(X); \mathbb{Q})\,,$$
where we write
$$H^*(\mathcal{M}^\Lambda; \mathbb{Q}) = H^*(\mathcal{M}^\Lambda; \mathbb{Q})^{\Gamma_r} \oplus H^*(\mathcal{M}^\Lambda; \mathbb{Q})^{\mathrm{var}}$$
as the direct sum of the Γ_r-invariant part and the non-invariant part, or *variant* part in the terminology of [**22**]. It follows from this that
$$P_t(\mathcal{M}) = P_t(\mathcal{M}^\Lambda)(1+t)^{2g}$$
if and only if Γ_r acts trivially on $H^*(\mathcal{M}^\Lambda; \mathbb{Q})$: in fact,

(12.2) $$P_t(\mathcal{M}^\Lambda)(1+t)^{2g} - P_t(\mathcal{M}) = P_t^{\mathrm{var}}(\mathcal{M}^\Lambda)(1+t)^{2g}\,,$$

where $P_t^{\mathrm{var}}(\mathcal{M}^\Lambda) = \sum t^i \dim(H^i(\mathcal{M}^\Lambda; \mathbb{Q})^{\mathrm{var}})$ is the Poincaré polynomial corresponding to the variant part of the cohomology.

2. Morse indices

The S^1-action on $\mathcal{M}$ restricts to the fixed determinant moduli space $\mathcal{M}^\Lambda$ and the Morse theory explained in Chapter 3 can be applied to this latter space. Thus, the restriction of f to $\mathcal{M}^\Lambda \subset \mathcal{M}$ gives a perfect Bott–Morse function. The characterization of the critical points of the Morse function (i.e., the fixed points of the S^1-action) and their stability given in Propositions 3.4 and 3.6 remains valid. Hence, for each critical submanifold $\mathcal{N} \subseteq \mathcal{M}$, there is a corresponding critical submanifold $\mathcal{N}^\Lambda \subseteq \mathcal{M}^\Lambda$ and the determinant map (12.1) restricts to give a fibration

(12.3) $$\det \colon \mathcal{N} \to \operatorname{Jac}^{\bar{\Delta}}(X)$$

with fibre over Λ equal to $\mathcal{N}^\Lambda$. Note that there is no need to map to $T^* \operatorname{Jac}^{\bar{\Delta}}(X)$ because for any parabolic complex variation of Hodge structure $(\bigoplus E_l, \Phi)$ we have $\operatorname{Tr} \Phi = 0$.

REMARK 12.3. We have a description of $\mathcal{N}$ as a fibred product $\mathcal{N}^\Lambda \times_{\Gamma_r} \operatorname{Jac}^0(X)$, analogous to the one given in Remark 12.2 for $\mathcal{M}$. Thus we also have an analogous description of the relation between the cohomology of $\mathcal{N}$ and that of $\mathcal{N}^\Lambda$.

The deformation theory of $\mathbf{E} = (E, \Phi)$ in the fixed determinant moduli space is governed by the complex
$$C_0^\bullet(\mathbf{E}): \quad \operatorname{ParEnd}_0(E) \xrightarrow{[-,\Phi]} \operatorname{SParEnd}_0(E) \otimes K(D)$$
$$f \mapsto (f \otimes 1)\Phi - \Phi f\,,$$

where the subscript 0 indicates trace zero (cf. Proposition 2.2). Now let $\mathbf{E} = (\bigoplus E_l, \Phi)$ be a fixed point of the S^1-action. In order to determine the weight spaces of the infinitesimal circle action on the tangent space we modify the subcomplexes $C^\bullet(\mathbf{E})_l$ defined in Section 2 of Chapter 3 to be subcomplexes $C_0^\bullet(\mathbf{E})_l$ of trace zero endomorphisms. Note that $C_0^\bullet(\mathbf{E})_l = C^\bullet(\mathbf{E})_l$ unless $l = 0$ or $l = -1$. The calculation of the Morse indices now proceeds analogously to the non-fixed determinant case of Chapter 3 and, in particular, we obtain the following result.

PROPOSITION 12.4. *Let the parabolic Higgs bundle $\mathbf{E} = (E, \Phi)$ represent a critical point of the restriction of f to $\mathcal{M}^\Lambda \subset \mathcal{M}$. Then the Morse index of f at this point is given by the formula of Proposition 3.11.*

PROOF. The Morse index of the restriction of f to $\mathcal{M}^\Lambda \subset \mathcal{M}$ equals the real dimension of the space $\bigoplus_{l>0} \mathbb{H}^1(C_0^\bullet(E, \Phi)_l)$. But, as pointed out above, $C_0^\bullet(\mathbf{E})_l = C^\bullet(\mathbf{E})_l$ for $l > 0$. This proves the proposition. Alternatively, we could have appealed to the invariance of f under the action of the Jacobian on $\mathcal{M}$ by tensor product. □

REMARK 12.5. The analogue of Theorem 3.14 also holds in the fixed determinant case, with an analogous proof.

3. Critical submanifolds of type $(1, 1, 1)$

In this section we describe the critical submanifolds of type $(1, 1, 1)$ and their contribution to the Poincaré polynomial in the fixed determinant case. We shall use the notations of Chapter 7. Note that the description given in Section 1 of Chapter 7 of the parabolic Higgs bundles which corresponds to critical points of type $(1, 1, 1)$ remains valid. Likewise, the characterization of stability given in Proposition 7.1 is the same. Thus, fixing d_1, m and ϖ, the fixed determinant critical submanifold is the fibre of the map det defined in (12.3):

$$\mathcal{N}^\Lambda_{(1,1,1)}(d_1, m, \varpi) = \det{}^{-1}(\Lambda) \ .$$

The description of the critical submanifolds now proceeds as in [16] (cf. also Hausel–Thaddeus [22] for the case of general rank r) to give us the following fixed determinant analogue of Proposition 7.2.

PROPOSITION 12.6. *The critical submanifold $\mathcal{N}^\Lambda_{(1,1,1)}(d_1, m, \varpi)$ is given by the pull-back diagram*

$$\begin{array}{ccc} \mathcal{N}^\Lambda_{(1,1,1)}(d_1, m, \varpi) & \longrightarrow & \mathrm{Jac}^{d_3}(X) \\ \downarrow & & \downarrow \\ S^{m_1}X \times S^{m_2}X & \longrightarrow & \mathrm{Jac}^{m_1+2m_2}(X) \ , \end{array}$$

where the vertical map on the left is $(L_1 \oplus L_2 \oplus L_3, \Phi_1, \Phi_2) \mapsto (\mathrm{div}(\Phi_1), \mathrm{div}(\Phi_2))$, the map in the bottom line is $(D_1, D_2) \mapsto \mathcal{O}(D_1 + 2D_2)$ and the vertical map on the right is $L_3 \mapsto \Lambda^{-1} \otimes L_3^3 \otimes K^3(3D - S_1 - 2S_2)$. Moreover, $\mathcal{N}^\Lambda_{(1,1,1)}(d_1, m, \varpi)$ is *non-empty if and only if*

$$3m > 2\Delta - F(\alpha, \varpi),$$
$$3d_1 > \Delta - G(\alpha, \varpi),$$
$$2d_1 - m \leq n - s_1 + 2g - 2,$$
$$2m - d_1 \leq \Delta + n - s_2 + 2g - 2,$$

where F and G were defined in (7.1).

PROOF. Given a parabolic Higgs bundle $(L_1 \oplus L_2 \oplus L_3, \Phi_1, \Phi_2)$ of type $(1,1,1)$, let M_i be the line bundle associated to the divisor $D_i = \text{div}(\Phi_i)$:

$$M_1 = L_1^{-1} \otimes L_2 \otimes K(D - S_1),$$
$$M_2 = L_2^{-1} \otimes L_3 \otimes K(D - S_2).$$

Then

(12.4) $$M_1 \otimes M_2^2 = \Lambda^{-1} \otimes L_3^3 \otimes K^3(3D - S_1 - 2S_2).$$

Conversely, let (d_1, m, ϖ) be such that $m_i \geq 0$ for $i = 1, 2$, take effective divisors D_i of degree m_i and define $M_i = \mathcal{O}(D_i)$. Then there is a solution L_3 to (12.4), determined up to the choice of a cube root of the trivial bundle. Once this choice is made, the isomorphism class of $(L_1 \oplus L_2 \oplus L_3, \Phi_1, \Phi_2)$ can be recovered from (D_1, D_2). Now, the first two inequalities of the statement of the proposition represent the stability condition for $(L_1 \oplus L_2 \oplus L_3, \Phi_1, \Phi_2)$ and the last two inequalities are equivalent to $m_i \geq 0$ for $i = 1, 2$. Thus we see that, for any (d_1, m, ϖ) satisfying these conditions, there is a non-empty critical submanifold, as described in the statement of the proposition. □

REMARK 12.7. Alternatively, we can parametrize the critical submanifolds by (m_1, m_2, ϖ). We then have $\mathcal{N}^\Lambda_{(1,1,1)}(m_1, m_2, \varpi) = \mathcal{N}^\Lambda_{(1,1,1)}(d_1, m, \varpi)$, which are non-empty if and only if

(12.5)
$$m_1 + 2m_2 < 6g - 6 + 3n - s_1 - 2s_2 + F(\alpha, \varpi),$$
$$2m_1 + m_2 < 6g - 6 + 3n - 2s_1 - s_2 + G(\alpha, \varpi),$$
$$m_1 \geq 0,$$
$$m_2 \geq 0,$$

and

(12.6) $$m_1 + 2m_2 + \Delta + s_1 + 2s_2 \equiv 0 \pmod{3}.$$

The conditions (12.5) are obtained by formulating the conditions of the preceding proposition in terms of m_1 and m_2, and the condition (12.6) must be added for it to be possible to solve (12.4) for L_3 (as pointed out in [**22**], this condition was overlooked in [**16**]).

As noted in Remark 12.3, the rational cohomology of $\mathcal{N}^\Lambda_{(1,1,1)}(d_1, m, \varpi)$ splits in an invariant part and a variant part, under the action of Γ_3:

$$H^*(\mathcal{N}^\Lambda_{(1,1,1)}(d_1, m, \varpi)) = H^*(\mathcal{N}^\Lambda_{(1,1,1)}(d_1, m, \varpi))^{\Gamma_3} \oplus H^*(\mathcal{N}^\Lambda_{(1,1,1)}(d_1, m, \varpi))^{\text{var}}.$$

PROPOSITION 12.8. *The invariant part of the cohomology of* $\mathcal{N}^\Lambda_{(1,1,1)}(d_1, m, \varpi)$ *is given by*

$$H^*(\mathcal{N}^\Lambda_{(1,1,1)}(d_1, m, \varpi))^{\Gamma_3} \cong H^*(S^{m_1} X \times S^{m_2} X).$$

The variant part of the cohomology is concentrated in degree $m_1 + m_2$ and has dimension

$$(3^{2g} - 1)\binom{2g-2}{m_1}\binom{2g-2}{m_2} .$$

PROOF. This is essentially [**16**, Proposition 3.11], cf. also [**22**]. □

Given this result, we can now find the contribution to the Poincaré polynomial of $\mathcal{M}^\Lambda$ coming from the invariant part of the critical submanifolds of type $(1,1,1)$.

PROPOSITION 12.9. *Under Assumption 7.6, the contribution of the invariant part of the cohomology of critical submanifolds of type $(1,1,1)$ to the Poincaré polynomial of $\mathcal{M}^\Lambda$ is*

$$P_t^{\Gamma_3}(\Lambda,(1,1,1)) = \underset{u^0 v^0}{\mathrm{Coeff}}\bigg((1 + 2u^2vt^2 + 2uv^2t^2 + u^3v^3t^4)^n \cdot$$
$$\cdot \frac{t^{2(4g-3+n)}}{u^{3n+6g-9+\Delta_0}v^{3n+6g-6-\Delta_0}} \cdot$$
$$\cdot \frac{(1 + u^2vt)^{2g}(1 + uv^2t)^{2g}}{(1-u^2v)(1-uv^2)(1-u^2vt^2)(1-uv^2t^2)(1-v^3t^2)(1-u^3t^2)}\bigg),$$

where $\Delta_0 \in \{1,2\}$ is the remainder modulo 3 of $\Delta = \deg(\Lambda)$.

PROOF. The proof proceeds exactly as in Chapter 7, except that we omit the factor $(1+t)^{2g}$ coming from the Jacobian (cf. Propositions 7.2, 12.4 and 12.8). □

It remains to find the contribution from the variant part of the critical submanifolds of type $(1,1,1)$.

PROPOSITION 12.10. *Under Assumption 7.6, the contribution of the variant part of the cohomology of critical submanifolds of type $(1,1,1)$ to the Poincaré polynomial of $\mathcal{M}^\Lambda$ is*

$$P_t^{\mathrm{var}}(\Lambda,(1,1,1)) = 2 \cdot 6^{n-1}(3^{2g} - 1)t^{12g-12+6n}(t+1)^{4g-4} .$$

PROOF. It is convenient to parametrize the critical submanifolds by (m_1, m_2, ϖ) as explained in Remark 12.7. In order to do the calculation, we therefore need to express the Morse index in terms of these invariants. Using Proposition 12.4 and Lemma 7.4 we obtain

$$\lambda_{(m_1,m_2,\varpi)} = 16g - 16 + 6n - 2m_1 - 2m_2 .$$

Hence, using Proposition 12.8, the contribution from the variant part of the cohomology of critical submanifolds of type $(1,1,1)$ for fixed ϖ is

(12.7) $$\sum t^{16g-16+6n-m_1-m_2}(3^{2g} - 1)\binom{2g-2}{m_1}\binom{2g-2}{m_2} ,$$

where the sum is over (m_1, m_2) satisfying the conditions (12.5) and (12.6). Note that the terms in the sum are only non-zero when $0 \le m_i \le 2g-2$. But Assumption 7.6 implies that the region defined by (12.5) contains all such (m_1, m_2). Therefore we can sum over all (m_1, m_2), subject to the condition (12.6). For this, let $\xi =$

$e^{2\pi i/3}$, then $\sum_{j=1}^{3} \xi^{j\nu}$ equals 3 if $\nu \equiv 0 \pmod{3}$ and zero otherwise. It follows that we can rewrite (12.7) as

$$\frac{1}{3} \sum_{m_1,m_2} \sum_{j=1}^{3} \xi^{j(m_1+2m_2+\Delta+s_1+2s_2)} t^{16g-16+6n-m_1-m_2} (3^{2g}-1) \binom{2g-2}{m_1} \binom{2g-2}{m_2}$$

$$= \frac{(3^{2g}-1)t^{16g-16+6n}}{3} \sum_{j=1}^{3} \xi^{j(\Delta+s_1+2s_2)}(1+t^{-1}\xi^j)^{2g-2}(1+t^{-1}\xi^{2j})^{2g-2}$$

$$= \frac{(3^{2g}-1)t^{12g-12+6n}}{3} \left((\xi^{\Delta+s_1+2s_2} + \xi^{2(\Delta+s_1+2s_2)})(t^2-t+1)^{2g-2} + (t+1)^{4g-4}\right).$$

It remains to do the sum over $\varpi \in (S_3)^n$. For this we use Table 7.1 to obtain

$$\sum_{\varpi} \xi^{\Delta+s_1+2s_2} = \sum_{\varpi} \xi^{2(\Delta+s_1+2s_2)} = 0.$$

Since the number of elements of $(S_3)^n$ is 6^n, we therefore obtain the result of the statement of the proposition. □

4. Parabolic triples of fixed determinant

Now we want to describe the moduli spaces of parabolic triples with fixed determinant, that we shall use in the following section to deal with the critical submanifolds of types $(1,2)$ and $(2,1)$. We follow the notations of Chapters 4 and 5. Fixing the topological and parabolic types of the triples, there is a determinant map on the moduli space $\mathcal{N}_\sigma$ of σ-stable parabolic triples,

$$\det : \mathcal{N}_\sigma \to \text{Jac}(X)$$
$$T = (E_1, E_2, \phi) \mapsto \det(E_1) \otimes \det(E_2).$$

We define the moduli space of σ-stable parabolic triples with fixed determinant Λ as

$$\mathcal{N}_\sigma^\Lambda = \det^{-1}(\Lambda).$$

In order to state the deformation theory of the parabolic triples with fixed determinant, we need to introduce the following subcomplex of $C^\bullet(T,T)$,

$$C_0^\bullet(T,T) : \big(\text{ParHom}(E_1, E_1) \oplus \text{ParHom}(E_2, E_2)\big)_0 \xrightarrow{c} \text{SParHom}(E_2, E_1(D)),$$

where $\big(\text{ParHom}(E_1, E_1) \oplus \text{ParHom}(E_2, E_2)\big)_0$ is defined as the kernel of the map

$$\text{ParHom}(E_1, E_1) \oplus \text{ParHom}(E_2, E_2) \to \mathcal{O},$$
$$(a_1, a_2) \mapsto \text{Tr}(a_1) + \text{Tr}(a_2).$$

We have the following result.

THEOREM 12.11. *Let $T = (E_1, E_2, \phi)$ be a σ-stable parabolic triple with determinant Λ.*

(i) *The Zariski tangent space at the point defined by T in the moduli space of stable triples with fixed determinant is isomorphic to $\mathbb{H}^1(C_0^\bullet(T,T))$.*

(ii) *If $\mathbb{H}^2(C_0^\bullet(T,T)) = 0$, then the moduli space of σ-stable parabolic triples with fixed determinant is smooth in a neighbourhood of the point defined by T.*

(iii) *If ϕ is injective or surjective then $T = (E_1, E_2, \phi)$ defines a smooth point in the moduli space $\mathcal{N}_\sigma^\Lambda$.*

PROOF. Items (i) and (ii) follow from Theorem 4.12. For (iii), let us define the complex $C_{\det}^\bullet : \mathcal{O} \to 0$. This complex is embedded in $C^\bullet(T,T)$ as

$$\begin{array}{ccccc} C_{\det}^\bullet & : & \mathcal{O} & \longrightarrow & 0 \\ \downarrow & & \downarrow & & \downarrow \\ C^\bullet(T,T) & : & \mathrm{ParHom}(E_1,E_1) \oplus \mathrm{ParHom}(E_2,E_2) & \longrightarrow & \mathrm{SParHom}(E_2, E_1(D)), \end{array}$$

where the left map is $\lambda \mapsto (\lambda\,\mathrm{Id}, \lambda\,\mathrm{Id})$. Then it is easy to see that we have a direct sum splitting of complexes as

(12.8) $$C^\bullet(T,T) = C_0^\bullet(T,T) \oplus C_{\det}^\bullet.$$

Now, if ϕ is injective or surjective, then the decomposition (12.8) gives that $0 = \mathbb{H}^2(C^\bullet(T,T)) = \mathbb{H}^2(C_0^\bullet(T,T)) \oplus \mathbb{H}^2(C_{\det}^\bullet)$, from where we get the result stated in (iii). □

In order to study the variation of the moduli spaces $\mathcal{N}_\sigma^\Lambda$ when moving σ, we follow the arguments of Chapter 5. We keep the notations of that section and work under Assumption 5.1. Consider a critical value σ_c. There is a determinant map

$$\begin{aligned} \det : B_{\sigma_c} = \mathcal{N}'_{\sigma_c} \times \mathcal{N}''_{\sigma_c} &\to \mathrm{Jac}(X), \\ (T', T'') &\mapsto \det T' \otimes \det T''. \end{aligned}$$

We introduce the following subspace of B_{σ_c},

(12.9) $$B_{\sigma_c}^\Lambda = \det^{-1}(\Lambda).$$

The flip loci in the moduli space of parabolic triples with fixed determinant are given by

$$\mathcal{S}_{\sigma_c^\pm}^\Lambda = \mathcal{S}_{\sigma_c^\pm} \cap \mathcal{N}_{\sigma_c^\pm}^\Lambda \subset \mathcal{N}_{\sigma_c^\pm}^\Lambda.$$

The description of $\mathcal{S}_{\sigma_c^\pm}^\Lambda$ follows the arguments of Section 2 of Chapter 5. We get the following.

PROPOSITION 12.12. (i) If $\mathbb{H}^2(C^\bullet(T',T')) = 0$ and $\mathbb{H}^2(C^\bullet(T'',T'')) = 0$ for every $(T',T'') \in B_{\sigma_c}^\Lambda$, then $B_{\sigma_c}^\Lambda$ is smooth.
 (ii) If $\mathbb{H}^2(C^\bullet(T'',T')) = 0$ and $\mathbb{H}^2(C^\bullet(T',T'')) = 0$ for every $(T',T'') \in B_{\sigma_c}^\Lambda$, then $\mathcal{S}_{\sigma_c^\pm}^\Lambda = \mathbb{P}\left(W^\pm|_{B_{\sigma_c}^\Lambda}\right)$, where $W^\pm|_{B_{\sigma_c}^\Lambda}$ is the restriction of $W^\pm \to B_{\sigma_c}$ to $B_{\sigma_c}^\Lambda$.

PROOF. The tangent space to $B_{\sigma_c}^\Lambda$ is given by the following subcomplex of the complex $C^\bullet(T',T') \oplus C^\bullet(T'',T'')$,

$$C_d^\bullet(T',T'') : \begin{pmatrix} \mathrm{ParHom}(E_1',E_1') \oplus \\ \oplus \mathrm{ParHom}(E_2',E_2') \oplus \\ \oplus \mathrm{ParHom}(E_1'',E_1'') \oplus \\ \oplus \mathrm{ParHom}(E_2'',E_2'') \end{pmatrix}_0 \longrightarrow \begin{array}{c} \mathrm{SParHom}(E_2',E_1'(D)) \oplus \\ \oplus \mathrm{SParHom}(E_2'',E_1''(D)) \end{array}$$

where the $C_d^0(T',T'')$ is the kernel of the map

$$\begin{array}{ccc} \mathrm{ParHom}(E_1',E_1') \oplus \mathrm{ParHom}(E_2',E_2') \oplus & \longrightarrow & \mathcal{O}, \\ \oplus \mathrm{ParHom}(E_1'',E_1'') \oplus \mathrm{ParHom}(E_2'',E_2'') & & \\ (a_1', a_2', a_1'', a_2'') & \mapsto & \begin{array}{c} \mathrm{Tr}(a_1') + \mathrm{Tr}(a_2') + \\ + \mathrm{Tr}(a_1'') + \mathrm{Tr}(a_2''). \end{array} \end{array}$$

Again there is a splitting of complexes

$$C^\bullet(T',T') \oplus C^\bullet(T'',T'') = C_d^\bullet(T',T'') \oplus C_{\det}^\bullet,$$

where $C^\bullet_{\det} \hookrightarrow C^\bullet(T', T') \oplus C^\bullet(T'', T'')$ is given by the map

$$\begin{array}{rl} \mathcal{O} \to & \mathrm{ParHom}(E'_1, E'_1) \oplus \mathrm{ParHom}(E'_2, E'_2) \oplus \\ & \oplus \mathrm{ParHom}(E''_1, E''_1) \oplus \mathrm{ParHom}(E''_2, E''_2), \\ \lambda \mapsto & (\lambda \,\mathrm{Id}, \lambda \,\mathrm{Id}, \lambda \,\mathrm{Id}, \lambda \,\mathrm{Id}). \end{array}$$

This proves that $\mathbb{H}^2(C^\bullet_d) = 0$ and hence that $B^\Lambda_{\sigma_c}$ is smooth. The second item follows from Proposition 5.9. □

PROPOSITION 12.13. *Assume that $\mathcal{N}^\Lambda_{\sigma_c^\pm}$ and $B^\Lambda_{\sigma_c}$ are smooth, and that it holds $\mathbb{H}^2(C^\bullet(T'', T')) = 0$ and $\mathbb{H}^2(C^\bullet(T', T'')) = 0$ for every $(T', T'') \in B^\Lambda_{\sigma_c}$. Let $\widetilde{\mathcal{N}}^\Lambda_{\sigma_c^\pm}$ be the blow-up of $\mathcal{N}^\Lambda_{\sigma_c^\pm}$ along $\mathcal{S}^\Lambda_{\sigma_c^\pm}$. Then*

$$\widetilde{\mathcal{N}}^\Lambda_{\sigma_c^+} \cong \widetilde{\mathcal{N}}^\Lambda_{\sigma_c^-}.$$

PROOF. The proof of Proposition 5.11 needs some slight modifications to the situation of fixed determinant. The complex $C^\bullet(\mathcal{T}, \mathcal{T})$ used to compute the tangent bundle to $\mathcal{N}_{\sigma_c^+}$ should be substituted by $C^\bullet_0(\mathcal{T}, \mathcal{T})$, which computes the tangent bundle to $\mathcal{N}^\Lambda_{\sigma_c^+}$. Likewise, the complex $C^\bullet(\mathcal{T}', \mathcal{T}') \oplus C^\bullet(\mathcal{T}'', \mathcal{T}'')$ should be substituted by the complex $C^\bullet_d(\mathcal{T}', \mathcal{T}'')$ introduced above, which deals with the tangent bundle to $B^\Lambda_{\sigma_c}$. Also, the piece

$$\mathrm{ParHom}_U(\mathcal{E}_1, \mathcal{E}_1) \oplus \mathrm{ParHom}_U(\mathcal{E}_2, \mathcal{E}_2)$$

in the complex computing the tangent bundle to $\mathbb{P}W^+$ must be substituted by the kernel $\big(\mathrm{ParHom}_U(\mathcal{E}_1, \mathcal{E}_1) \oplus \mathrm{ParHom}_U(\mathcal{E}_2, \mathcal{E}_2)\big)_0$ of

$$\begin{array}{rl} \mathrm{ParHom}_U(\mathcal{E}_1, \mathcal{E}_1) \oplus \mathrm{ParHom}_U(\mathcal{E}_2, \mathcal{E}_2) & \to \quad \mathcal{O}, \\ (a_1, a_2) & \mapsto \quad \mathrm{Tr}(a_1) + \mathrm{Tr}(a_2). \end{array}$$

Taking this into account, we reach the conclusion that the normal bundle to $\mathcal{S}^\Lambda_{\sigma_c^\pm}$ in $\mathcal{N}^\Lambda_{\sigma_c^\pm}$ is isomorphic to $(p^* W^\mp \otimes \mathcal{O}_{\mathbb{P}W^\pm}(-1))|_{B^\Lambda_{\sigma_c}}$.

Using this last fact, the arguments of Proposition 5.12 carry over verbatim to prove the stated isomorphism. □

Finally we apply Proposition 12.13 to compute the Poincaré polynomial of the moduli spaces of σ-stable triples with fixed determinant for the case of ranks $r_1 = 2$ and $r_2 = 1$. We follow the notations of Chapter 6. The description of the flip loci also holds in this situation. Let $\sigma_c > \sigma_m$ be a critical value. Then we have the following equality of Poincaré polynomials:

$$(12.10) \qquad P_t\big(\mathcal{N}^\Lambda_{\sigma_c^-}\big) - P_t\big(\mathcal{N}^\Lambda_{\sigma_c^+}\big) = P_t\big(\mathbb{P}\left(W^-_{\sigma_c}|_{B^\Lambda_{\sigma_c}}\right)\big) - P_t\big(\mathbb{P}\left(W^+_{\sigma_c}|_{B^\Lambda_{\sigma_c}}\right)\big),$$

where $W^\pm_{\sigma_c}|_{B^\Lambda_{\sigma_c}}$ is a projective fibration over $B^\Lambda_{\sigma_c}$ with fibres projective spaces of dimension $w^\pm_{\sigma_c} - 1$. But $B^\Lambda_{\sigma_c} = \det^{-1}(\Lambda)$ where the determinant map is

$$\begin{array}{rl} \mathcal{N}'_{\sigma_c} \times \mathcal{N}''_{\sigma_c} = \mathrm{Jac}^{d_M} X \times (\mathrm{Jac}^{d_2} X \times S^N X) & \to \quad \mathrm{Jac}(X), \\ (M, L, Z) & \mapsto \quad M \otimes L \otimes L(Z). \end{array}$$

Therefore we have an isomorphism

$$B^\Lambda_{\sigma_c} \cong \mathrm{Jac}\, X \times S^N X.$$

The arguments of the proof of Theorem 6.4 now give the following result.

THEOREM 12.14. *Let $\sigma > \sigma_m$ be a non-critical value. For any $\varepsilon = \{\varepsilon(p)\}_{p \in D}$, $\varepsilon(p) \in \{1, 2\}$, let s_1, s_2, s_3 and $\bar{d}_M$ be defined as in Theorem 6.4. Then*

$$P_t(\mathcal{N}_\sigma^\Lambda) = \sum_\varepsilon \underset{x^0}{\mathrm{Coeff}} \left(\frac{(1+t)^{2g}(1+xt)^{2g}t^{2d_1-2d_2+2s_2+2s_3-2\bar{d}_M}x^{\bar{d}_M}}{(1-t^2)(1-x)(1-xt^2)(1-t^{-2}x)x^{d_1-d_2+s_1}} \right.$$
$$\left. - \frac{(1+t)^{2g}(1+xt)^{2g}t^{-2d_1+2g-2+2n-2s_3+4\bar{d}_M}x^{\bar{d}_M}}{(1-t^2)(1-x)(1-xt^2)(1-t^4x)x^{d_1-d_2+s_1}} \right). $$

□

5. Critical submanifolds of type $(1, 2)$ and $(2, 1)$

First consider the critical submanifolds of type $(1, 2)$. We shall use the notations of Chapter 8. Note that the description given in Section 1 of Chapter 8 of the parabolic Higgs bundles which corresponds to critical points of type $(1, 2)$ remains valid. Thus, fixing d_1 and ϖ, the fixed determinant critical submanifold is the fibre of the map det defined in (12.3):

$$\mathcal{N}_{(1,2)}^\Lambda(d_1, \varpi) = \det{}^{-1}(\Lambda) .$$

The characterization of stability given in Proposition 4.6 tells us that $\mathcal{N}_{(1,2)}^\Lambda(d_1, \varpi)$ is isomorphic to the moduli space of σ-stable triples (of the appropriate topological and parabolic type) with fixed determinant $\Lambda \otimes K^2$, as considered in Section 4, for $\sigma = 2g - 2$. Therefore Theorem 12.14 and the computations of Chapter 8 give the following.

PROPOSITION 12.15. *Under Assumption 7.6, the contribution of the critical submanifolds of type $(1, 2)$ to the Poincaré polynomial of $\mathcal{M}^\Lambda$ is*

$$P_t(\Lambda, (1, 2)) = $$
$$= \underset{x^0}{\mathrm{Coeff}} \left(\frac{(1+t)^{2g}(1+xt)^{2g}t^{8g-8+2n}x^{5-2g-\Delta_0-n}(1+2t^2+2t^2x+t^4x)^n}{(1-t^2)(1-x)(1-xt^2)(1-t^{-2}x)(1-t^2x^2)} \right.$$
$$\left. - \frac{(1+t)^{2g}(1+xt)^{2g}t^{6g+6-4\Delta_0}x^{5-2g-\Delta_0-n}(1+2t^2+2t^4x+t^6x)^n}{(1-t^2)(1-x)(1-xt^2)(1-t^4x)(1-t^8x^2)} \right),$$

where $\Delta_0 \in \{1, 2\}$ is the remainder modulo 3 of $\Delta = \deg(\Lambda)$. □

Now consider the critical submanifolds of type $(2, 1)$. Use the notations of Chapter 9. Fixing d_1 and ϖ, the fixed determinant critical submanifold is the fibre of the map det defined in (12.3):

$$\mathcal{N}_{(2,1)}^\Lambda(d_1, \varpi) = \det{}^{-1}(\Lambda) .$$

Lemma 9.1 holds in this situation, telling us that $\mathcal{N}_{(2,1)}^\Lambda(d_1, \varpi)$ is isomorphic to the moduli space of σ-stable triples of type $(1, 2)$, with appropriate degrees and weights, with fixed determinant $\Lambda^{-1} \otimes K^{-1}(-3D)$, and for $\sigma = 2g - 2$. The computations of Chapter 9 together with Theorem 12.14 yield the following.

PROPOSITION 12.16. *Under Assumption 7.6, the contribution of the critical submanifolds of type $(2,1)$ to the Poincaré polynomial of $\mathcal{M}^\Lambda$ is*

$$P_t(\Lambda,(2,1)) =$$
$$= \operatorname*{Coeff}_{x^0} \left(\frac{(1+t)^{2g}(1+xt)^{2g}t^{8g-8+2n}x^{2-2g+\Delta_0-n}(1+2t^2+2t^2x+t^4x)^n}{(1-t^2)(1-x)(1-xt^2)(1-t^{-2}x)(1-t^2x^2)} \right.$$
$$\left. - \frac{(1+t)^{2g}(1+xt)^{2g}t^{6g-6+4\Delta_0}x^{2-2g+\Delta_0-n}(1+2t^2+2t^4x+t^6x)^n}{(1-t^2)(1-x)(1-xt^2)(1-t^4x)(1-t^8x^2)} \right),$$

where $\Delta_0 \in \{1,2\}$ is the remainder modulo 3 of $\Delta = \deg(\Lambda)$. □

REMARK 12.17. The Poincaré polynomials of the critical submanifolds of type $(1,2)$ and $(2,1)$ for fixed and non-fixed determinant differ by a factor of $(1+t)^{2g}$, coming from the Jacobian. Hence Γ_3 acts trivially on the rational cohomology of the fixed determinant critical submanifolds of type $(1,2)$ and $(2,1)$ (cf. Remarks 12.2 and 12.3). The triviality of the action can also be seen directly from our description of the critical submanifolds as moduli spaces of triples, by using the flips picture and arguing as in the proof of [**22**, Lemma 10.5].

6. Critical submanifolds of type (3)

The critical points of type (3) are just the stable parabolic bundles and hence the corresponding critical submanifold is the moduli space of stable parabolic bundles of fixed determinant Λ. As pointed out in Remark 10.1, the fixed determinant case was the one studied by Holla [**25**], and we obtain the Poincaré polynonial of the fixed determinant moduli space by dividing the formula of Proposition 10.4 by $(1+t)^{2g}$. Thus we have the following.

PROPOSITION 12.18. *Under Assumption 7.6, the Poincaré polynomial of the moduli space of stable parabolic bundles of rank 3 of fixed determinant Λ is given by*

$$P_t(\Lambda, 3) = \frac{(1+2t^2+2t^4+t^6)^{n-1}}{(1-t^2)^3(1-t^4)} \cdot \Big((1+t^3)^{2g}(1+t^5)^{2g}$$
$$+ (1+t)^{4g}(1+t^2+t^4)t^{6g-2} - (1+t)^{2g}(1+t^3)^{2g}(1+t^2)^2 t^{4g-2}\Big).$$

□

REMARK 12.19. Note, in particular, that Γ_3 acts trivially on the rational cohomology of the moduli space of stable parabolic bundles of rank 3 of fixed determinant (cf. Remarks 12.2 and 12.3).

7. Betti numbers of the fixed determinant moduli space

Finally we put everything together to obtain the Poincaré polynomial of the moduli space of rank three parabolic Higgs bundles of fixed determinant Λ.

THEOREM 12.20. *Let $\mathcal{M}^\Lambda$ be the moduli space of rank three parabolic Higgs bundles of fixed determinant Λ and some fixed weights, over a connected, smooth projective complex algebraic curve of genus g. If the weights are generic (in the sense* that there are no properly semistable parabolic Higgs bundles), then the Poincaré polynomial of $\mathcal{M}^\Lambda$ is given by

$$P_t(\mathcal{M}^\Lambda) = \underset{u^0 v^0}{\mathrm{Coeff}}\left((1+2u^2vt^2+2uv^2t^2+u^3v^3t^4)^n \cdot \right.$$

$$\left. \cdot \frac{t^{2(4g-3+n)}(1+u^2vt)^{2g}(1+uv^2t)^{2g}}{u^{3n+6g-8}v^{3n+6g-7}(1-u^2v)(1-uv^2)(1-u^2vt^2)(1-uv^2t^2)(1-v^3t^2)(1-u^3t^2)}\right)$$

$$+ \underset{x^0}{\mathrm{Coeff}}\left(\frac{(1+t)^{2g}(1+xt)^{2g}t^{6g-6}x^{2-2g-n}}{(1-t^2)(1-x)(1-xt^2)}\cdot \right.$$

$$\left. \cdot \left(\frac{t^{2g-2+2n}(x+x^2)(1+2t^2+2t^2x+t^4x)^n}{(1-t^{-2}x)(1-t^2x^2)} - \frac{(t^4x+t^8x^2)(1+2t^2+2t^4x+t^6x)^n}{(1-t^4x)(1-t^8x^2)}\right)\right)$$

$$+ \frac{(1+2t^2+2t^4+t^6)^{n-1}}{(1-t^2)^3(1-t^4)} \cdot \left((1+t^3)^{2g}(1+t^5)^{2g} + \right.$$

$$\left. + (1+t)^{4g}(1+t^2+t^4)t^{6g-2} - (1+t)^{2g}(1+t^3)^{2g}(1+t^2)^2t^{4g-2}\right)$$

$$+ 2 \cdot 6^{n-1}(3^{2g}-1)t^{12g-12+6n}(t+1)^{4g-4} \ .$$

$\square$

PROOF. The theorem follows by an argument analogous to the proof of Theorem 11.1, but now using the contributions from the fixed determinant critical submanifolds given in Propositions 12.9, 12.10, 12.15, 12.16 and 12.18. Also, Proposition 12.1 takes the place of Proposition 2.1. $\square$

COROLLARY 12.21. *The Euler characteristic of the moduli space of parabolic Higgs bundles with fixed determinant Λ is*

$$\chi(\mathcal{M}^\Lambda) = 0 \ .$$

PROOF. This could be shown by substituting $t = -1$ in the formula of Theorem 12.20. But it is, in fact, easier to note that the Euler characteristic of the moduli space equals the sum of the Euler characteristics of the critical submanifolds. Our description of these shows that they all have zero Euler characteristic. Hence the only potentially non-zero contribution comes from the invariant part of the cohomology of the critical submanifolds of type $(1,1,1)$, given in Proposition 12.8. From MacDonald's formula [29] we have $\chi(S^{m_i}X) = (-1)^{m_i}\binom{2g-2}{m_i}$ and hence

$$\chi(\mathcal{M}^\Lambda) = \sum (-1)^{m_1+m_2}\binom{2g-2}{m_1}\binom{2g-2}{m_2} \ ,$$

where the sum is over all (m_1, m_2, ϖ) satisfying the conditions (12.5) and (12.6). This is essentially the calculation of the proof of Proposition 12.10, with t substituted by -1, and gives zero. $\square$

Finally, we can identify the variant part of the cohomology of $\mathcal{M}^\Lambda$ under the action of Γ_3—this should be relevant for proving the rank 3 parabolic version, stated in [21], of the mirror symmetry theorem of Hausel–Thaddeus [22].

THEOREM 12.22. *The variant part of the rational cohomology of $\mathcal{M}^\Lambda$ has Poincaré polynomial*

$$P_t^{\mathrm{var}}(\mathcal{M}^\Lambda) = 2 \cdot 6^{n-1}(3^{2g}-1)t^{12g-12+6n}(t+1)^{4g-4} \ .$$

PROOF. As we have seen in Remarks 12.17 and 12.19, the critical submanifolds of type $(1,2)$, $(2,1)$ and (3) do not contribute to the variant cohomology. Hence, under Assumption 7.6, the variant Poincaré polynomial $P_t^{\mathrm{var}}(\mathcal{M}^\Lambda)$ equals the contribution coming from critical submanifolds of type $(1,1,1)$, given in Proposition 12.10. But, as we saw in (12.2),
$$P_t(\mathcal{M}^\Lambda)(1+t)^{2g} - P_t(\mathcal{M}) = P_t^{\mathrm{var}}(\mathcal{M}^\Lambda)(1+t)^{2g},$$
and we know from Propositions 2.1 and 12.1 that the left hand side is independent of the choice of Δ and parabolic weights made in Assumption 7.6. Hence the right hand side is also independent of this choice. This finishes the proof. □

Bibliography

[1] L. Álvarez-Cónsul and O. García-Prada, Dimensional reduction, SL(2, $\mathbb{C}$)-equivariant bundles and stable holomorphic chains, *Internat. J. Math.* **12** (2001) 159–201.

[2] M. F. Atiyah and R. Bott, The Yang-Mills equations over Riemann surfaces, *Philos. Trans. Roy. Soc. London Ser. A* **308** (1982) 523–615.

[3] O. Biquard, Fibrés paraboliques stables et connexions singulières plates, *Bull. Soc. Math. Fr.* **119** (1991) 231–257.

[4] O. Biquard and O. García-Prada, Parabolic vortex equations and instantons of infinite energy, *J. Geom. Physics* **21** (1997) 238–254.

[5] I. Biswas and S. Ramanan, An infinitesimal study of the moduli of Hitchin pairs, *J. London Math. Soc.* (2) **49** (1994) 219–231.

[6] H.U. Boden and K. Yokogawa, Moduli spaces of parabolic Higgs bundles and parabolic $K(D)$ pairs over smooth curves. I, *Internat. J. Math.* **7** (1996) 573–598.

[7] F. Bottacin, Symplectic geometry on moduli spaces of stable pairs, *Ann. Sci. Éc. Norm. Supér., IV. Sér.* **28** (1995) 391–433.

[8] S.B. Bradlow and O. García-Prada, Stable triples, equivariant bundles and dimensional reduction, *Math. Ann.* **304** (1996) 225–252.

[9] S.B. Bradlow, O. García-Prada and P.B. Gothen, Moduli spaces of holomorphic triples over compact Riemann surfaces, *Math. Ann.* **328** (2004) 299–351.

[10] S.A. Cherkis and A. Kapustin, Hyper-Kähler metrics from periodic monopoles, *Physical Review D*, **65** (2002) 084015.

[11] K. Corlette, Flat G-bundles with canonical metrics, *J. Differential Geom.* **28** (1988) 361–382.

[12] S.K. Donaldson, Twisted harmonic maps and the self-duality equations, *Proc. London Math. Soc.* **55** (1987) 127–131.

[13] T. Frankel, Fixed points and torsion on Kähler manifolds, *Ann. Math.* **70** (1959) 1–8.

[14] V. Ginzburg, The global nilpotent variety is Lagrangian, *Duke Math. J.* **109** (2001) 511–519.

[15] P. Griffiths and J. Harris, *Principles of Algebraic Geometry*, John Wiley & Sons (1978).

[16] P.B. Gothen, The Betti numbers of the moduli space of stable rank 3 Higgs bundles on a Riemann surface, *Internat. J. Math.* **5** (1994) 861–875.

[17] P.B. Gothen, Components of spaces of representations and stable triples, *Topology* **40** (2001) 823–850.

[18] P.B. Gothen and A.D. King, Homological algebra of twisted quiver bundles, J. London Math. Soc. (2) **71** (2005) 85–99.

[19] T. Hausel, Compactification of moduli of Higgs bundles, *J. Reine Angew. Math.* **503** (1998) 169–192.

[20] T. Hausel, Mirror symmetry and Langlands duality in the non-Abelian Hodge theory of a curve, in *Geometric Methods in Algebra and Number Theory*, Progress in Mathematics, Vol. 235 F. Bogomolov,Y. Tschinkel (Eds.) 2005.

[21] T. Hausel and M. Thaddeus, Examples of mirror partners arising from integrable systems, *C. R. Acad. Sci. Paris Sér. 1* **333** (2001) 313–318.

[22] T. Hausel and M. Thaddeus, Mirror symmetry, Langlands duality, and the Hitchin system, *Invent. Math.* **153** (2003) 197–229.

[23] N.J. Hitchin, The self-duality equations on a Riemann surface, *Proc. London Math. Soc.* (3) **55** (1987) 59–126.

[24] N.J. Hitchin, Stable bundles and integrable systems, *Duke Math. J.* **54** (1987) 91–114.

[25] Y.I. Holla, Poincaré polynomial of the moduli space of parabolic bundles, *Proc. Indian Acad. Sci* **110** (2000) 233–261.

[26] H. Konno, Construction of the moduli space of stable parabolic Higgs bundles on a Riemann surface, *J. Math. Soc. Japan* **45** (1993) 253–276.

[27] H. Lange, Universal families of extensions, *J. Algebra* **83** (1983) 101–112.

[28] G. Laumon, Un analogue global du cône nilpotent, *Duke Math. J.* **57**(1988) 647–671.

[29] I.G. Macdonald, Symmetric products of an algebraic curve, *Topology* **1** (1962) 319–343.

[30] E.Markman, Spectral curves and integrable systems, *Comp. Math.* **93** (1994) 255–290.

[31] M. Maruyama and K. Yokogawa, Moduli of parabolic stable sheaves, *Math. Ann.* **293** (1992) 77–99.

[32] H. Nakajima, Hyper-Kähler structures on moduli spaces of parabolic Higgs bundles on Riemann surfaces, *Moduli of vector bundles.* Marcel Dekker. Lect. Notes Pure Appl. Math. 179 (1996) 199–208. Edited by Maruyama and Masaki.

[33] B. Nasatyr and B. Steer, Orbifold Riemann surfaces and the Yang–Mills–Higgs equations, *Ann. Scuola Norm. Sup. Pisa Cl. Sci. (4)* **22** (1995) 595–643.

[34] N. Nitsure, Cohomology of the moduli space of parabolic vector bundles, *Proc. Indian Acad. Sci* **95** (1986) 61–77.

[35] C.T. Simpson, Harmonic bundles on noncompact curves, *J. Amer. Math. Soc.* **3** (1990) 713–770.

[36] C.T. Simpson, Higgs bundles and local systems *Inst. Hautes Études Sci. Publ. Math.* **75** (1992) 5–95.

[37] M. Thaddeus, Stable pairs, linear systems and the Verlinde formula, *Invent. Math.* **117** (1994) 317–353.

[38] M. Thaddeus, Variation of moduli of parabolic Higgs bundles, *J. Reine Angew. Math.* **547** (2002) 1–14.

[39] K. Yokogawa, Compactification of moduli of parabolic sheaves and moduli of parabolic Higgs sheaves, *J. Math. Kyoto Univ.* **33** (1993) 451–504.

[40] K. Yokogawa, Infinitesimal deformation of parabolic Higgs sheaves, *Internat. J. Math.* **6** (1995) 125–148.

Editorial Information

To be published in the *Memoirs*, a paper must be correct, new, nontrivial, and significant. Further, it must be well written and of interest to a substantial number of mathematicians. Piecemeal results, such as an inconclusive step toward an unproved major theorem or a minor variation on a known result, are in general not acceptable for publication.

Papers appearing in *Memoirs* are generally at least 80 and not more than 200 published pages in length. Papers less than 80 or more than 200 published pages require the approval of the Managing Editor of the Transactions/Memoirs Editorial Board.

As of February 28, 2007, the backlog for this journal was approximately 15 volumes. This estimate is the result of dividing the number of manuscripts for this journal in the Providence office that have not yet gone to the printer on the above date by the average number of monographs per volume over the previous twelve months, reduced by the number of volumes published in four months (the time necessary for preparing a volume for the printer). (There are 6 volumes per year, each usually containing at least 4 numbers.)

A Consent to Publish and Copyright Agreement is required before a paper will be published in the *Memoirs*. After a paper is accepted for publication, the Providence office will send a Consent to Publish and Copyright Agreement to all authors of the paper. By submitting a paper to the *Memoirs*, authors certify that the results have not been submitted to nor are they under consideration for publication by another journal, conference proceedings, or similar publication.

Information for Authors

Memoirs are printed from camera copy fully prepared by the author. This means that the finished book will look exactly like the copy submitted.

Initial submission. The AMS uses Centralized Manuscript Processing for initial submissions. Authors should submit a PDF file using the Initial Manuscript Submission form found at www.ams.org/cgi-bin/peertrack/submission.pl, or send one copy of the manuscript to the following address: Centralized Manuscript Processing, MEMOIRS OF THE AMS, 201 Charles Street, Providence, RI 02904-2294 USA. If a paper copy is being forwarded to the AMS, indicate that it is for it Memoirs and include the name of the corresponding author, contact information such as email address or mailing address, and the name of an appropriate Editor to review the paper (see the list of Editors below).

The paper must contain a *descriptive title* and an *abstract* that summarizes the article in language suitable for workers in the general field (algebra, analysis, etc.). The *descriptive title* should be short, but informative; useless or vague phrases such as "some remarks about" or "concerning" should be avoided. The *abstract* should be at least one complete sentence, and at most 300 words. Included with the footnotes to the paper should be the 2000 *Mathematics Subject Classification* representing the primary and secondary subjects of the article. The classifications are accessible from www.ams.org/msc/. The list of classifications is also available in print starting with the 1999 annual index of *Mathematical Reviews*. The Mathematics Subject Classification footnote may be followed by a list of *key words and phrases* describing the subject matter of the article and taken from it. Journal abbreviations used in bibliographies are listed in the latest *Mathematical Reviews* annual index. The series abbreviations are also accessible from www.ams.org/publications/. To help in preparing and verifying references, the AMS offers MR Lookup, a Reference Tool for Linking, at www.ams.org/mrlookup/.

Electronically prepared manuscripts. The AMS encourages electronically prepared manuscripts, with a strong preference for $\mathcal{A}_{\mathcal{M}}\mathcal{S}$-LaTeX. To this end, the Society has prepared $\mathcal{A}_{\mathcal{M}}\mathcal{S}$-LaTeX author packages for each AMS publication. Author packages include instructions for preparing electronic manuscripts, samples, and a style file that generates the particular design specifications of that publication series. Though $\mathcal{AMS}$-LaTeX is the highly preferred format of TeX, author packages are also available in $\mathcal{AMS}$-TeX.

Authors may retrieve an author package from the AMS website starting from `www.ams.org/tex/` or via FTP to `ftp.ams.org` (login as `anonymous`, enter username as password, and type `cd pub/author-info`). The *AMS Author Handbook* and the *Instruction Manual* are available in PDF format following the author packages link from `www.ams.org/tex/`. The author package can also be obtained free of charge by sending email to `tech-support@ams.org` (Internet) or from the Publication Division, American Mathematical Society, 201 Charles St., Providence, RI 02904-2294, USA. When requesting an author package, please specify $\mathcal{AMS}$-LaTeX or $\mathcal{AMS}$-TeX and the publication in which your paper will appear. Please be sure to include your complete mailing address.

After acceptance. The final version of the electronic file should be sent to the Providence office (this includes any TeX source file, any graphics files, and the DVI or PostScript file) immediately after the paper has been accepted for publication.

Before sending the source file, be sure you have proofread your paper carefully. The files you send must be the EXACT files used to generate the proof copy that was accepted for publication. For all publications, authors are required to send a printed copy of their paper, which exactly matches the copy approved for publication, along with any graphics that will appear in the paper.

Accepted electronically prepared files can be submitted via the web at `www.ams.org/submit-book-journal/`, sent via FTP, or sent on CD-Rom or diskette to the Electronic Prepress Department, American Mathematical Society, 201 Charles Street, Providence, RI 02904-2294 USA. TeX source files, DVI files, and PostScript files can be transferred over the Internet by FTP to the Internet node `ftp.ams.org` (130.44.1.100). When sending a manuscript electronically via CD-Rom or diskette, please be sure to include a message identifying the paper as a Memoir.

Electronically prepared manuscripts can also be sent via email to `pub-submit@ams.org` (Internet). In order to send files via email, they must be encoded properly. (DVI files are binary and PostScript files tend to be very large.)

Electronic graphics. Comprehensive instructions on preparing graphics are available at `www.ams.org/jourhtml/`. A few of the major requirements are given here.

Submit files for graphics as EPS (Encapsulated PostScript) files. This includes graphics originated via a graphics application as well as scanned photographs or other computer-generated images. If this is not possible, TIFF files are acceptable as long as they can be opened in Adobe Photoshop or Illustrator. No matter what method was used to produce the graphic, it is necessary to provide a paper copy to the AMS.

Authors using graphics packages for the creation of electronic art should also avoid the use of any lines thinner than 0.5 points in width. Many graphics packages allow the user to specify a "hairline" for a very thin line. Hairlines often look acceptable when proofed on a typical laser printer. However, when produced on a high-resolution laser imagesetter, hairlines become nearly invisible and will be lost entirely in the final printing process.

Screens should be set to values between 15% and 85%. Screens which fall outside of this range are too light or too dark to print correctly. Variations of screens within a graphic should be no less than 10%.

Inquiries. Any inquiries concerning a paper that has been accepted for publication should be sent to `memo-query@ams.org` or directly to the Electronic Prepress Department, American Mathematical Society, 201 Charles St., Providence, RI 02904-2294 USA.

Editors

This journal is designed particularly for long research papers, normally at least 80 pages in length, and groups of cognate papers in pure and applied mathematics. Papers intended for publication in the *Memoirs* should be addressed to one of the following editors. The AMS uses Centralized Manuscript Processing for initial submissions to AMS journals. Authors should follow instructions listed on the Initial Submission page found at www.ams.org/memo/memosubmit.html.

Algebra to ALEXANDER KLESHCHEV, Department of Mathematics, University of Oregon, Eugene, OR 97403-1222; email: ams@noether.uoregon.edu

Algebra and its application to MINA TEICHER, Emmy Noether Research Institute for Mathematics, Bar-Ilan University, Ramat-Gan 52900, Israel; email: teicher@macs.biu.ac.il

Algebraic geometry to DAN ABRAMOVICH, Department of Mathematics, Brown University, Box 1917, Providence, RI 02912; email: amsedit@math.brown.edu

Algebraic number theory to V. KUMAR MURTY, Department of Mathematics, University of Toronto, 100 St. George Street, Toronto, ON M5S 1A1, Canada; email: murty@math.toronto.edu

Algebraic topology to ALEJANDRO ADEM, Department of Mathematics, University of British Columbia, Room 121, 1984 Mathematics Road, Vancouver, British Columbia, Canada V6T 1Z2; email: adem@math.ubc.ca

Combinatorics to JOHN R. STEMBRIDGE, Department of Mathematics, University of Michigan, Ann Arbor, Michigan 48109-1109; email: FRS@umich.edu

Complex analysis and harmonic analysis to ALEXANDER NAGEL, Department of Mathematics, University of Wisconsin, 480 Lincoln Drive, Madison, WI 53706-1313; email: nagel@math.wisc.edu

Differential geometry and global analysis to LISA C. JEFFREY, Department of Mathematics, University of Toronto, 100 St. George St., Toronto, ON Canada M5S 3G3; email: jeffrey@math.toronto.edu

Dynamical systems and ergodic theory to AMIE WILKINSON, Department of Mathematics, Northwestern University, 2033 Sheridan Road, Evanston, IL 60208-2730; email: transactions@math.northwestern.edu

Functional analysis and operator algebras to DIMITRI SHLYAKHTENKO, Department of Mathematics, University of California, Los Angeles, CA 90095; email: shlyakht@math.ucla.edu

Geometric analysis to WILLIAM P. MINICOZZI II, Department of Mathematics, Johns Hopkins University, 3400 N. Charles St., Baltimore, MD 21218; email: trans@math.jhu.edu

Geometric analysis to MLADEN BESTVINA, Department of Mathematics, University of Utah, 155 South 1400 East, JWB 233, Salt Lake City, Utah 84112-0090; email: bestvina@math.utah.edu

Harmonic analysis, representation theory, and Lie theory to ROBERT J. STANTON, Department of Mathematics, The Ohio State University, 231 West 18th Avenue, Columbus, OH 43210-1174; email: stanton@math.ohio-state.edu

Logic to STEFFEN LEMPP, Department of Mathematics, University of Wisconsin, 480 Lincoln Drive, Madison, Wisconsin 53706-1388; email: lempp@math.wisc.edu

Partial differential equations to GUSTAVO PONCE, Department of Mathematics, South Hall, Room 6607, University of California, Santa Barbara, CA 93106; email: ponce@math.ucsb.edu

Partial differential equations and dynamical systems to PETER POLACIK, School of Mathematics, University of Minnesota, Minneapolis, MN 55455; email: polacik@math.umn.edu

Probability and statistics to KRZYSZTOF BURDZY, Department of Mathematics, University of Washington, Box 354350, Seattle, Washington 98195-4350; email: burdzy@math.washington.edu

Real analysis and partial differential equations to DANIEL TATARU, Department of Mathematics, University of California, Berkeley, Berkeley, CA 94720; email: tataru@math.berkeley.edu

All other communications to the editors should be addressed to the Managing Editor, ROBERT GURALNICK, Department of Mathematics, University of Southern California, Los Angeles, CA 90089-1113; email: guralnic@math.usc.edu.

Titles in This Series

879 **O. García-Prada, P. B. Gothen, and V. Muñoz,** Betti numbers of the moduli space of rank 3 parabolic Higgs bundles, 2007

878 **Alessandra Celletti and Luigi Chierchia,** KAM stability and celestial mechanics, 2007

877 **María J. Carro, José A. Raposo, and Javier Soria,** Recent developments in the theory of Lorentz spaces and weighted inequalities, 2007

876 **Gabriel Debs and Jean Saint Raymond,** Borel liftings of Borel sets: Some decidable and undecidable statements, 2007

875 **C. Krattenthaler and T. Rivoal,** Hypergéométrie et fonction zêta de Riemann, 2007

874 **Sonia Natale,** Semisolvability of semisimple Hopf algebras of low dimension, 2007

873 **A. J. Duncan,** Exponential genus problems in one-relator products of groups, 2007

872 **Anthony V. Geramita, Tadahito Harima, Juan C. Migliore, and Yong Su Shin,** The Hilbert function of a level algebra, 2007

871 **Pascal Auscher,** On necessary and sufficient conditions for L^p-estimates of Riesz transforms associated to elliptic operators on $\mathbb{R}^n$ and related estimates, 2007

870 **Takuro Mochizuki,** Asymptotic behaviour of tame harmonic bundles and an application to pure twistor D-modules, Part 2, 2007

869 **Takuro Mochizuki,** Asymptotic behaviour of tame harmonic bundles and an application to pure twistor D-modules, Part 1, 2007

868 **Gelu Popescu,** Entropy and multivariable interpolation, 2006

867 **Vilmos Totik,** Metric properties of harmonic measures, 2006

866 **William Craig,** Semigroups underlying first-order logic, 2006

865 **Nathanial P. Brown,** Invariant means and finite representation theory of $C*$-algebras, 2006

864 **John M. Lee,** Fredholm operators and Einstein metrics on conformally compact manifolds, 2006

863 **M. Lübke and A. Teleman,** The Universal Kobayashi-Hitchin correspondence on Hermitian manifolds, 2006

862 **Alberto Canonaco,** The Beilinson complex and canonical rings of irregular surfaces, 2006

861 **Leon A. Takhtajan and Lee-Peng Teo,** Weil-Petersson metric on the universal Teichmüller space, 2006

860 **Thomas M. Fiore,** Pseudo limits, biadjoints and pseudo algebras: Categorical foundations of conformal field theory, 2006

859 **N. Arcozzi, R. Rochberg, and E. Sawyer,** Carleson measures and interpolating sequences for Besov spaces on complex balls, 2006

858 **Enrico Valdinoci, Berardino Sciunzi, and Vasile Ovidiu Savin,** Flat level set regularity of p-Laplace phase transitions, 2006

857 **Donatella Danielli, Nocola Garofalo, and Duy-Minh Nhieu,** Non-doubling Ahlfors measures, perimeter measures, and the characterization of the trace spaces of Sobolev functions in Carnot-Carathéodory spaces, 2006

856 **Vladimir Bolotnikov and Harry Dym,** On boundary interpolation for matrix valued Schur functions, 2006

855 **Yevgenia Kashina, Yorck Sommerhäuser, and Yongchang Zhu,** On higher Frobenius-Schur indicators, 2006

854 **Noam Greenberg,** The role of true finiteness in the admissible recursively enumerable degrees, 2006

853 **Joachim Krieger,** Stability of spherically symmetric wave maps, 2006

852 **Viorel Barbu, Irena Lasiecka, and Roberto Triggiani,** Tangential boundary stabilization of Navier-Stokes equations, 2006

TITLES IN THIS SERIES

851 **Jie Wu,** On maps from loop suspensions to loop spaces and the shuffle relations on the Cohen groups, 2006

850 **Siegfried Echterhoff, S. Kaliszewski, John Quigg, and Iain Raeburn,** A categorical approach to imprimitivity theorems for C^*-dynamical systems, 2006

849 **Katsuhiko Kuribayashi, Mamoru Mimura, and Tetsu Nishimoto,** Twisted tensor products related to the cohomology of the classifying spaces of loop groups, 2006

848 **Bob Oliver,** Equivalences of classifying spaces completed at the prime two, 2006

847 **Eric T. Sawyer and Richard L. Wheeden,** Hölder continuity of weak solutions to subelliptic equations with rough coefficients, 2006

846 **Victor Beresnevich, Detta Dickinson, and Sanju Velani,** Measure theoretic laws for lim–sup sets, 2006

845 **Ehud Friedgut, Vojtech Rödl, Andrzej Ruciński, and Prasad V. Tetali,** A Sharp threshold for random graphs with a monochromatic triangle in every edge coloring, 2006

844 **Amadeu Delshams, Rafael de la Llave, and Tere M. Seara,** A geometric mechanism for diffusion in Hamiltonian systems overcoming the large gap problem: Heuristics and rigorous verification on a model, 2006

843 **Denis V. Osin,** Relatively hyperbolic groups: Intrinsic geometry, algebraic properties, and algorithmic problems, 2006

842 **David P. Blecher and Vrej Zarikian,** The calculus of one-sided M-ideals and multipliers in operator spaces, 2006

841 **Enrique Artal Bartolo, Pierrette Cassou-Noguès, Ignacio Luengo, and Alejandro Melle Hernández,** Quasi-ordinary power series and their zeta functions, 2005

840 **Sławomir Kołodziej,** The complex Monge-Ampère equation and pluripotential theory, 2005

839 **Mihai Ciucu,** A random tiling model for two dimensional electrostatics, 2005

838 **V. Jurdjevic,** Integrable Hamiltonian systems on complex Lie groups, 2005

837 **Joseph A. Ball and Victor Vinnikov,** Lax-Phillips scattering and conservative linear systems: A Cuntz-algebra multidimensional setting, 2005

836 **H. G. Dales and A. T.-M. Lau,** The second duals of Beurling algebras, 2005

835 **Kiyoshi Igusa,** Higher complex torsion and the framing principle, 2005

834 **Ken'ichi Ohshika,** Kleinian groups which are limits of geometrically finite groups, 2005

833 **Greg Hjorth and Alexander S. Kechris,** Rigidity theorems for actions of product groups and countable Borel equivalence relations, 2005

832 **Lee Klingler and Lawrence S. Levy,** Representation type of commutative Noetherian rings III: Global wildness and tameness, 2005

831 **K. R. Goodearl and F. Wehrung,** The complete dimension theory of partially ordered systems with equivalence and orthogonality, 2005

830 **Jason Fulman, Peter M. Neumann, and Cheryl E. Praeger,** A generating function approach to the enumeration of matrices in classical groups over finite fields, 2005

829 **S. G. Bobkov and B. Zegarlinski,** Entropy bounds and isoperimetry, 2005

828 **Joel Berman and Paweł M. Idziak,** Generative complexity in algebra, 2005

827 **Trevor A. Welsh,** Fermionic expressions for minimal model Virasoro characters, 2005

826 **Guy Métivier and Kevin Zumbrun,** Large viscous boundary layers for noncharacteristic nonlinear hyperbolic problems, 2005

For a complete list of titles in this series, visit the
AMS Bookstore at **www.ams.org/bookstore/**.

WITHDRAWN